29	HCO^+	Aldehydes
30	$CH_2NH_2^+$	Amines
31	$H_2C=\overset{+}{O}H$ CH_3O^+	Alcohols Methyl esters
43	$C_3H_7^+$ CH_3CO^+	Propyl ion Acetyl groups
47	CH_3S^+	Sulfides
49	CH_2Cl^+	Chloro compounds
55	$C_4H_7^+$	Alkyl groups
57	$C_4H_9^+$ $C_2H_5CO^+$	Alkyl groups Acylium ion
58	$(H_2C\overset{H}{\overset{\diagup}{C}}OCH_3)^{+}$	Ketones with a γ-hydrogen
59	$(COOCH_3)^+$	Methyl esters
61	$CH_3\overset{OH}{\underset{}{C}}=O^+H$	Esters of high molecular weight alcohols
70		Pyrrolidines
74	$(CH_2\overset{OH}{\underset{}{C}}-OCH_3)^{+}$	Methyl esters with a γ-hydrogen
77	$C_6H_5^+$	Aromatics
78	$C_5H_4N^+$	Pyridines and alkyl pyrroles
80		Pyrroles
80(82)	HBr^{+}	Bromo compounds
81		Furans
		Aliphatic chain with two double bonds
83	$C_6H_{11}^+$ $CHCl_2^+$	Cyclohexanes or hexenes Chloro compounds

85		Tetrahydropyranyl ethers
88	$CH_3OCO-CH$ $\overset{}{\underset{+NH_2}{}}$	Amino acid esters
91	$C_7H_7^+$	Aromatic hydrocarbons with side chains
92	$C_7H_8^+$	Benzyl compounds with a γ-hydrogen
95	CO^+	Furyl-CO-X
97		Alkyl thiophenes
99		Etylene ketals of cyclic compounds (steroids)
104	$C_8H_8^{+}$	Alkyl aromatics
105	$C_6H_5CO^+$ $C_8H_9^+$	Benzoyl compounds Aromatic hydrocarbons
106	$CH_2=\overset{+}{\underset{NH_2}{}}$	Amino benzyl
107	$C_7H_7O^+$	Phenolic hydrocarbons
117	$C_9H_9^+$	Styrenes
128	HI^+	Iodo compounds
130		Indoles
131	CO^+	Cinnamates
149	$\overset{CO}{\underset{CO}{}}OH^+$	Dialkyl phthalates (rearrangement)

Common Fragment Ions

MASS SPECTROMETRY DESK REFERENCE

FIRST EDITION

O. DAVID SPARKMAN

Global View Publishing
Pittsburgh, Pennsylvania

Mass Spectrometry Desk Reference

By: O. David Sparkman
E-mail: ods@compuserve.com

Global View Publishing
Post Office Box 111384
Pittsburgh, PA 15238-9998
Telephone: 1-888-LCMS.COM or
1-412-967-6881
Fax: 1-412-963-6882

Ordering Information

Individual Sales.   Global View Publications are available through most bookstores. They can also be ordered direct from Global View Publishing at the address above or logging onto "The LC/MS Book Store" at www.LCMS.COM.

Printed in the United States of America

Publisher's Cataloging-in-Publication
(Provided by Quality Books, Inc.)

Sparkman, O. David (Orrin David), 1942-
        Mass spectrometry desk reference / O. David
Sparkman. – 1st ed.
        p. cm.
        Includes bibliographical references and index.
        LCCN: 00-100995
        ISBN: 0-9660813-2-3

        1. Mass spectrometry—Handbooks, manuals, etc.
I. Title.

QD96.M3S67 2000   543.0873
                    QBI00-218

This book is dedicated to my partner, my best friend, my companion—my wife, Joan;

and

to my partner in mass spectrometry whom many of you have often heard me refer to—our dog, Maggie.

Table of Contents

FOREWARD   ix

PREFACE   xi

ACKNOWLEDGMENTS   xiii

CONVENTIONS USED IN THIS BOOK   xv

CORRECT AND INCORRECT TERMS   1

 Introduction...1
 Data..3
 Data Acquisition Techniques and Ionization 7
 Ions...24
 Mass 34
 Some Other Important Definitions 39
 Some Other Terminology To Avoid 45
 Instruments...46
 Terms Associated with Double-Focusing
  Mass Spectrometers 49
 Terms Associated with Time-of-Flight
  Mass Spectrometers 51
 Use of Abbreviations 53
 Components of a Measurement 56
 Formulas and Equations...57
 Types of Elements And Electrons 58
 Terms Associated with Computerized
  Spectral Matching 59
 Ion Detection..63
 Referenced Citations in Documents 65
 References 66

BIBLIOGRAPHY   71

INDEX   101

FOREWARD

I am pleased to recommend and comment on this publication by my friend and colleague, O. David Sparkman. *Mass Spec Desk Reference* gives an excellent overview to, as well as specific details on, the language and scientific communication media used in the active world of mass spectrometry. Over the years, I have gained great respect for David's scientific integrity and curiosity as I have worked closely with him in developing several nationally recognized short courses on mass spectrometry for the American Chemical Society. David's attention to detail and his organizational skills, learned in part through his training as an analytical chemist, are evident in this document. The reader will benefit from David's Zen-like but pragmatic involvement in mass spectrometry, ranging from instrument rep to product manager to software developer to lecturer to general consultant. In the ACS courses, the novice in mass spectrometry has enjoyed David's allegories (e.g., "electrons and snakes like to travel in pairs" in explaining the driving force in homolytic bond fission, or "take off like a scalded dog" in describing acceleration of ions from an ion source). Some of this 'Texan straight-talk and humor' can be found herein.

J. Throck Watson
Professor, Departments of Biochemistry and Chemistry
Michigan State University
East Lansing, Michigan
January 2000

PREFACE

The analytical technique of mass spectrometry generates mass spectra regardless of the sample-introduction technique, the method of ion formation, or the way ions are separated. When a molecule is ionized, a characteristic ion, representing the intact molecule and/or a group of ions of different masses that represent fragments of the ionized molecule, is formed. When these ions are separated, the plot of their relative abundance versus the mass-to-charge ratio (m/z) of each ion constitutes a mass spectrum. Learning to identify a molecule from its mass spectrum is much easier than using any other type of spectral information. The mass spectrum shows the mass of the molecule and the masses of its pieces. Mass spectrometry offers more information about an analyte from less sample than any other technique. Mass spectrometry is also the most accurate technique for the determination of mass. The only disadvantage of mass spectrometry compared to other techniques is that, usually, the sample is consumed; however, so little sample is required, it is inconsequential.

Mass spectrometry had its origin in the works of the English physicist, Joseph John Thomson (1856–1940), who won the 1906 Nobel Prize in physics for the discovery of the electron. Thomson's work was further developed by Francis William Aston (1877–1945), another English physicist, a student of Thomson, and also a Nobel Prize winner (chemistry, 1922); and by Arthur Jeffrey Dempster (1886–1950), a Canadian-American physicist at the University of Chicago, who worked independently of Thomson and Aston. Aston built the first mass spectrometer based on experimental instrumentation developed by Thomson. Mass spectrometry developed as a distinct field during the period of 1911–1925 based on the efforts of these three outstanding scientists. Today, mass spectrometry has become one of the most widely used analytical techniques for the monitoring of respiratory gases to the surface analysis of construction materials to the sequencing of complex DNA molecules. Advances in techniques of ion formation and separation have stripped away the volatility and limited mass requirements of the analyte that were present in the formative years through the 1970s. Now that the technology has advanced to the point that mass spectrometrists "…can make elephants fly," there is no applications area of analytical chemistry that remains untouched by mass spectrometry.

A large amount of mass spectral data of small molecules comes about as a result of gas chromatography/mass spectrometry (GC/MS). GC/MS data are usually taken under electron ionization (EI)[1] conditions at about 70 eV. Some EI mass spectral data result from introduction of pure samples via a direct-insertion probe, and some by means of particle beam liquid chromatography/mass spectrometry. Most all of the spectral libraries and compilations are of EI spectra taken at 70 eV. A fundamental requirement of EI mass spectrometry is that it must be possible to put the analyte molecule in the gas phase under reduced pressure conditions. This requirement is not true of analytes that

---

[1] EI can refer to both electron ionization and electron impact ionization. The term "electron impact" was replaced with electron ionization after the abbreviation "CI" was adopted for chemical ionization. Electron impact is considered to be archaic.

are analyzed using one of the desorption/ionization techniques, which are often used for the analysis of macromolecules [e.g., liquid chromatography electrospray (ES), matrix-assisted laser desorption/ionization (MALDI) mass spectrometry, or fast atom bombardment (FAB)]. However, regardless of the method of ion formation or the size of the ionized molecule, fragments produced will always result from the loss of a logical grouping of atoms.

Because mass spectral data obtained by sample introduction through a chromatograph (gas or liquid) are a collection of mass spectra taken during the chromatographic separation of mixtures, they can be displayed as mass chromatograms (a plot of scan number versus the intensity of a single mass-to-charge ratio) or reconstructed total-ion-current chromatograms (RTICC),[2] both of which can be used for precise quantitation. This added dimension is a decisive advantage of mass spectrometry over other techniques and can be used to verify the validity of data and remove analyte contaminants. In the absence of a chromatograph as the sample-introduction device, it is especially important to be able to distinguish interpretable mass spectral peaks from background and decomposition peaks.

Spectra obtained using the so-called "soft-ionization" techniques of desorption and chemical ionization (DI and CI) can result in protonated and other adduct molecules that have a charge. The fragmentation of these ions are interpreted using the same general principles as used in the interpretation of EI spectra. Since these soft-ionization techniques involve a less-harsh treatment of the analyte, a molecular ion (or representation of it) is much more likely to be present in the spectrum. Large nonvolatile molecules can be analyzed with mass spectrometry. The development of techniques such as ES (and its variants) for liquid chromatography/mass spectrometry (LC/MS) has greatly aided the analysis of mixtures of nonvolatile analytes in complex biological and environmental matrices. New ionization techniques such as MALDI, coupled with mass spectrometers not limited by a maximum *m/z* value (the time-of-flight mass spectrometer), have made the analysis of carbohydrates, nucleic acids, peptides, and proteins a reality.

All of this means more people in more diverse disciplines will be using mass spectral data. Due to the lack of training (because mass spectrometry was not pertinent to their field of study at the time of their initial education), these people need tools such as provided by this book to help them in getting the most from the data and technique. As you begin or continue your study of mass spectrometry, try to develop your own tools so that your skills become second nature—like riding a bicycle.

---

[2] Many data systems call this TIC. TIC is used to reduce the amount of space required for the label.

ACKNOWLEDGMENTS

The **Correct and Incorrect Terms** section of this book has been in the preparation stage for almost 10 years. While the GC/MS trainer for the Varian CSB worldwide sales organization, it became obvious that it was necessary to originate such a document as a way to prevent people from (1) using "self-CI" to describe two separate phenomena that resulted in spectral anomalies (space-charge effects resulting from ion overload, and ion/molecule reactions caused primarily by odd-electron fragment ions protonating analyte molecules) in the quadrupole ion-trap mass spectrometer, and (2) using "amu" as a synonym for mass-to-charge ratio or as a symbol for atomic mass unit. Over the years, the document grew from its original four 8½ × 11 pages to somewhat less than its present size based on issues raised in the various American Chemical Society mass spectrometry short courses that I teach. In conversation (both verbally and electronically), reference would be made to this document; then, as a result, requests for copies would be made. Some of these requesters encouraged me to publish this document because of their perceived significance of the material.

In March of 1998, Dominic Desiderio, one of the co-editors of *Mass Spectrometry Reviews*, contacted me and said he had heard of the document and wanted to know if it could be submitted for possible publication. In April of that same year, my first draft was submitted. Because of the potential impact and some of my contradictions to what had been published by the American Society for Mass Spectrometry (ASMS) and the International Union of Pure and Applied Chemistry (IUPAC), Dominic elected to send the manuscript to more reviewers than normal. Based on personal comments that were received from some of those selected to review the draft, it was obvious that many, if not all, of the editors of major mass spectrometry journals had been selected as reviewers. All of the reviews were very thorough and had many helpful comments and suggestions as well as a few "who-do-you-think-you-are" statements that were made about my disagreements with the ASMS and IUPAC publications.

On April 2, 1999, a revised manuscript was submitted to *Mass Spectrometry Reviews* that addressed many of the issues raised by the reviewers and corrected a number of technical inaccuracies pointed out in the reviews. On May 1, 1999, a letter was sent to me that was jointly signed by the co-editors of *Mass Spectrometry Reviews* (Dominic M. Desiderio and Nico M. M. Nibbering), stating their concurrence with the "...consistent consensus among all of those reviewers that *Mass Spectrometry Reviews* might not be the appropriate publication in which to publish an article on correct and incorrect terms"—an opinion with which I cannot disagree.

The quality of the information, some additions and deletions, and the presentation itself is largely due to the comments of those reviewers who did a complete and thorough job in their reviews. Not one of the returned reviews was accomplished in a short period nor without considerable thought. Although I still do not have a consensus with all of the reviewers on all of the terms and their definitions, the article, "Correct and Incorrect Terms for Mass Spectrometry," has had the benefit of the peer-review process whose purpose is not only to evaluate the technical content but also to provide the author with the insight and wisdom of the reviewer, thereby resulting in a superior product. The co-editors of *Mass Spectrometry Reviews* and their selected reviewers deserve my deepest gratitude. This acknowledgment is in no way meant to imply that any of these people has expressed an endorsement of this particular work.

In addition, I would like to express my appreciation to my two colleagues with whom I teach the American Chemical Society short courses for their continual comments, inputs, and inspiration—J. Throck Watson and Frederick E. Klink. And last but by far not least, I owe my greatest appreciation for this book to my wife, Joan A. Sparkman, who has been the copy editor for each and every one of the hundreds of revised and expanded drafts.

O. David Sparkman

CONVENTIONS USED IN THIS BOOK

Some of the terms defined as **Correct** or **Incorrect** in this book are exceptions to the definitions found in the *Current IUPAC Recommendations* (Todd, 1995) and the *ASMS Guidelines* (Price, 1991). These terms are noted with an asterisk (*). This book is not meant to supersede compilations of official terms, but should be used as a clarification tool. It is important to remember that the aim is to communicate with a minimal amount of confusion.

In some cases where a term is listed as **Incorrect**, the text immediately following the term describes the incorrect usage. Text following this statement of incorrect usage will contain a correct definition and/or usage of the term. There may be circumstances where this term should be used.

Example:

Incorrect: **API** – when used as an abbreviation for atmospheric pressure chemical ionization (APCI). APCI was originally called atmospheric pressure ionization and abbreviated as API. This use is archaic and causes confusion.

In other cases, a term listed as **Incorrect** should not be used regardless of the circumstance.

Example:

Incorrect: **protonated molecular ion** – when used to describe a molecule that has been protonated. This term implies that the molecular ion (a positive-charge species) has reacted with a proton (another positive-charge species). For this reaction to happen, the two species would have to come in contact with one another. This event is unlikely because like charges repel one another.

Listings in each segment of the **Bibliography** section of this book are in chronological order.

CORRECT AND INCORRECT TERMS
FOR
MASS SPECTROMETRY

INTRODUCTION

The audience for mass spectrometry information is wider than it has ever been. Many members of this audience do not have English as their first language; many do not have extensive chemistry or scientific backgrounds (i.e., attorneys, physicians, engineers, etc.). Therefore, in working with various forms of mass spectrometry, it is important that the correct scientific terminology is used to describe the hardware, data, and technique. Unfortunately, the field of mass spectrometry does not follow a strict adherence to the use of SI units. Even the guidelines for terminology set forth by the International Union of Pure and Applied Chemistry (IUPAC) and the American Society for Mass Spectrometry (ASMS) contain esoteric neologisms that have become the accepted standards, contradictions, and typographical errors. This document is a guideline of correct and incorrect terms used in print and in oral presentations on the techniques and instrumentation of mass spectrometry and hyphenated mass spectral techniques such as gas or liquid chromatography/mass spectrometry (GC/MS or LC/MS). The primary purpose of this document is to draw attention to the need for standardized nomenclature so that all cultures (social and technical) can assimilate the scientific information that is being presented in English. Another purpose of this presentation is to collect terms that are often used with mass spectrometry (but that are not specific to such data or techniques) and define them along with mass spectrometry terms so that newcomers to the field can have a single reference.

Over the past three decades, there have been several compilations of terms for use with mass spectral techniques (Beynon, 1977; 1981; Karasek, Clement, 1988; Price, 1991; Todd, 1995; de Hoffmann, 1996; McLafferty, Tureček, 1993). Developing an accurate list of correct terms is difficult from two standpoints: 1) the ever-changing and advancing technology that requires new terms to communicate results and ideas; 2) the problem that many of the terms in mass spectrometry are not really being defined, and that authors have to be put in the position that they "know the definition." A good example of mass spectrometry nomenclature confusions are the definitions of nominal mass and monoisotopic mass. Three different sources define the nominal mass of an element in three different ways: 1) as the integer mass of each of its isotopes (Biemann, 1967); 2) as the integer mass of the lowest mass naturally occurring stable isotope of an element (Watson, 1997); 3) as the integer mass of the most abundant naturally occurring stable isotope—the correct definition (Bursey, 1971). Rather than defining monoisotopic mass as being based on the exact mass of the most abundant naturally occurring stable isotope of an element, another source defines this term as an integer mass value (de Hoffman, 1996).

The two references most often cited in this presentation of mass spectrometry terms are the *Current IUPAC Recommendations* (Todd, 1995) and the *ASMS Guidelines* (Price, 1991). The terms listed as "incorrect" are some of the more flagrant examples of problems in terminology. Some terms are listed as incorrect in order to eliminate the use of two different terms to describe the same thing, whereas other terms have been specifically designated as "no longer recommended" by the *Current IUPAC Recommendations*. Some of the various sources on terminology are not always in agreement with each other; and because this author, along with others, is not in complete agreement with the cited references that are presented here, a personal judgment will have to be made in case of a conflict. Some of the material contained in this document is clearly the preference of the author (e.g., restricting the use of magnetic sector to single-focusing instruments that use a magnetic field; the use of collisionally activated dissociation (CAD) in preference to collision-induced dissociation (CID); the incorrectness of the use of parent and daughter to describe ions, etc.). To include definitions such as those for atom, ion, etc. and terms used in organic chemistry such as heterolytic and homolytic cleavages may appear patronizing, but is not the intent. Definitions of these terms are necessary for a single source of information. This glossary is not intended to be a tutorial in mass spectrometry, but rather a collection of terms encountered in mass spectrometry.

Although this document is written as a guide to authors, it is intended more as a tool for people who deal with mass spectrometry and its data and only have a minimal knowledge of the field.

This glossary omits definitions that pertain to data systems, vacuum systems, and various mass spectrometry/mass spectrometry (MS/MS) techniques performed with double-focusing mass spectrometers. For information relating to vacuum systems, refer to the American Vacuum Institute, the Beynon report of the ASMS Nomenclature Workshop (Beynon, 1981), or the *Current IUPAC Recommendations*. The Beynon report is also a good reference for the definitions of various terms that are associated with MS/MS in double-focusing mass spectrometers. The MS/MS terminology in the Beynon report can also be found in the *ASMS Guidelines* and the *Current IUPAC Recommendations*. Data-system terminology is very specific to the products that are produced by the various mass spectrometer manufacturers and the computer industry. The data-system terminology that appears in the *Current IUPAC Recommendations* and the *ASMS Guidelines* is dated because it revolves around the minicomputer. Care should be taken in the use of terms that relate to data that are the creation of instrument manufacturers. A good source for references for data-system hardware and operating systems is the manufacturer of the computer. In the more recent years (ca. 1990), the problem of custom computers and operating systems for mass spectrometry has been relieved by the existence and proliferation of modern microcomputers.

There are terms that are more relevant to mass spectral instrumentation research and ion physics that appear in the *Current IUPAC Recommendations* and the *ASMS Guidelines*, but do not appear in this document. Therefore, both of these publications are important in the definition of mass spectral terms. This document is more oriented toward the applications of mass spectrometry.

One other area that is only peripherally addressed here is the use of hieroglyphs in mass spectrometry communications. Lehmann (Lehmann, 1997) has proposed a set of pictograms that could be added to the graphic display of mass spectral data and would allow a viewer to have an instant awareness of how the data were obtained and under what conditions. Although this convention has yet to be widely embraced, it offers a great deal of potential as do the graphics proposed for MS/MS acquisition modes (de Hoffmann, 1996).

DATA

Correct: **Ions** are found inside mass spectrometers, and **peaks** are found on paper or on the data record. The word **intensity** is used with respect to the height of a peak or to the strength of an ion beam. The word **abundance** is used to describe the number of ions in the mass spectrometer. Peaks are often displayed as straight lines in digitized mass spectral data. Mass spectral peaks can also be displayed as profiles with width as well as height. In GC/MS or LC/MS, care must be taken to clearly differentiate between **mass spectral peaks** and **chromatographic peaks**. The most intense peak in a displayed mass spectrum is the **base peak**.

Incorrect: **line** – when used to describe a mass spectral peak. This term has recently become popular in some circles but is incorrect, and does not refer to the actual digital plots that are often observed in a mass spectrum. Spectral lines are seen on a photographic plate in emission spectrography used for the elemental analysis of metals or in mass spectrographs—not in mass spectrometry.

Correct: **fragmentation pattern** – the result of the decomposition of a precursor ion to produce a series of product ions with specific abundances. This pattern is displayed as a mass spectrum.

Correct: **mass chromatogram** – describes the chromatographic data obtained from the plot of the intensity at a single mass-to-charge ratio (*m/z*) value or a range of *m/z* values that is a subset of the total range of each acquired spectrum versus the spectrum number. In the event that the sample is introduced into the mass spectrometer by some process other than a chromatographic one, the word **chromatogram** should be replaced with **profile** (e.g., an evaporation profile when the sample is evaporated (or sublimed) from a solid probe). In the case of capillary electrophoresis (CE), **chromatogram** is replaced with **pherogram**.

Incorrect: **reconstructed ion chromatogram (RIC)**, **extracted-ion-current (EIC)** or **reconstructed extracted-ion-current (REIC) chromatogram (EICC** or **REICC)** – when used as synonyms for mass chromatograms. All four of these terms have been used and are self-explanatory; however, confusion is best avoided when there is a single word or phase used to describe a specific type of data.

Correct: **mass spectrum** (singular) and **mass spectra** (plural) – the digital (each mass spectral peak represented by a single line) or analog (**profile data**) plot of the intensities observed at each acquired mass-to-charge ratio (see **Figures 1A** and **1B**). These data are often normalized relative to the most intense peak—the **base peak**. The mass spectrum can be displayed in a normalized or non-normalized table of *m/z* values and intensities. A mass spectrum can also be the presentation of data where the *m/z* values have been converted to mass by multiplying the *m/z* value by the value for *z*. The term **mass spectrum** was first used by Francis William Aston (English Nobel Laureate in chemistry, 1922, 1877–1945; one of the founding fathers of mass spectrometry) in 1920 (Kiser, 1965).

Correct: **monoisotopic mass spectrum** – a mass spectrum with peaks that represent the principle isotopes of the atoms that compose each ion. This presentation is usually a mass spectrum containing only nominal *m/z* value peaks representing each ion (molecular and fragment).

Incorrect: **MS/MS spectrum** or **MS^n spectrum** – when used to describe the mass spectrum that results from the controlled dissociation of a precursor ion or a series of 1st, 2nd, etc. generation product ions (various mass spectrometry/mass spectrometry techniques). An "MS/MS mass spectrum of n-butylbenzene" does not convey the same message as the "product-ion mass spectrum of the n-butylbenzene *m/z* 134 ion." Product-ion mass spectra often do not exhibit isotope peaks because a precursor ion of a single *m/z* value was selected.

Correct: **profile mass spectral data** – describes a mass spectrum where each mass spectral peak is displayed with a height and a width. The height at any point on the peak is a representation of the ion abundance. Each point along the width of the peak represents an ion of the same *m/z* value as that at the horizontal center of the peak but with a different energy and/or ions of different *m/z* values that are not resolved (see **Figure 1A**).

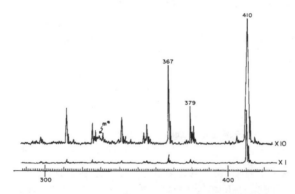

Figure 1A. Profile mass spectral data.

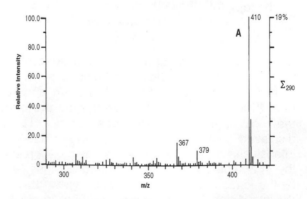

Figure 1B. Digitized mass spectral data.

Incorrect: **scan** – when used to refer to a mass spectrum. Scan is not correct for mass spectra that are acquired with a time-of-flight mass spectrometer (TOF-MS) or from a Fourier transform ion-cyclotron resonance mass spectrometer (FTICR-MS). The magnetic-sector mass spectrometer acquires data by scanning [ramping] the magnetic field strength. The quadrupole ion-trap mass spectrometer acquires data by scanning the amplitude of a fixed-frequency radio frequency (RF) voltage. The transmission-quadrupole mass spectrometer acquires data by holding the ratio of an RF amplitude to that of a direct current (DC) voltage constant, while increasing both. Although **scan** is correct for these latter three instruments, there is nothing "scanned" in a TOF-MS or an FTICR-MS; therefore, the word **scan** can lead to confusion in TOF mass spectrometry (TOFMS) or FTICR mass spectrometry (FTICRMS) and should not be used.

Correct: **reconstructed total-ion-current (RTIC) chromatogram (RTICC)** – defines a chromatographic plot prepared (reconstructed) from a consecutively recorded array of the sum of intensities of all the peaks in each spectrum versus the spectrum number (which is a function of a time domain). In the case of samples introduced into the mass spectrometer by some means other than a chromatographic technique, the plot would be a **reconstructed total-ion-current profile**. The important word in this phrase is **reconstructed**.

Note: **total-ion-current (TIC) chromatogram (TICC)** – a chromatographic output obtained when the total ion current is monitored in "real time" in the ion source. This type of output was used before the development of data systems to determine when to acquire a mass spectrum. The total ion current in the ion source is not monitored in most modern mass spectrometers, and the term was usually associated with magnetic-sector instruments of older vintages. Some manufacturers of GC/MS instrumentation incorrectly use total-ion chromatogram (TIC: abbreviation is in conflict with that used for total ion current) to refer to the **reconstructed total-ion-current chromatogram (RTICC)**.

Correct: **SIM mass chromatogram** – a chromatographic display that results from a **selected-ion-monitoring (SIM)** analysis.

Incorrect: **SIM plot** – when used to describe a mass chromatogram. The generation of a mass chromatogram is a different process than the use of the selected-ion-monitoring technique.

Incorrect: **total-ion chromatogram** – when used to describe an RTICC. It is possible to confuse the total-ion chromatogram used as a synonym for RTICC with the result obtained in ion chromatography, or it suggests the direct recording of the total ion current (TIC).

Correct: **total ion current (TIC)** – sometimes used to refer to the sum of all the intensities in a mass spectrum. TIC is synonymous with total ion abundance.

Correct: **mass discrimination** – the ability of an *m/z* analyzer to transmit or detect ions of one *m/z* value more efficiently than another. An instrument can exhibit either **high-** or **low-mass discrimination**. Mass discrimination can be due to instrument design or the contamination condition of the instrument.   Transmission-quadrupole mass spectrometers exhibit high-mass discrimination that is compensated for with either a **Brubaker prefilter**, which is a set of RF-only poles attached to the front of the filter (Brubaker, 1968), or a **Turner–Kruger ion-optics lens**, which is a lens that terminates inside the quadrupole field (Barnett, 1971).

Correct: **spectral skewing** – used to describe the phenomenon of changes in relative intensities of mass spectral peaks due to the changes in concentration of the analyte in the ion source as the chromatographic component elutes.   This phenomenon is not observed in ion-trap (quadrupole or magnetic) or time-of-flight (TOF) mass spectrometers. The TOF-MS records all ions sent down the flight tube in a single pulse. The ion-trap mass spectrometer records all the ions that have been stored in the trap.

Care must be taken not to confuse spectral skewing with mass discrimination. An example of spectral skewing is shown in **Figure 2**.

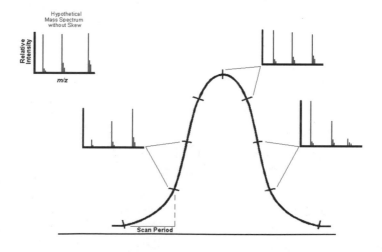

Figure 2. Illustration of spectral skewing.

DATA ACQUISITION TECHNIQUES AND IONIZATION

Correct: **atmospheric pressure ionization** (**API**) – causes ions to be in the gas phase at atmospheric pressure as opposed to the reduced pressure that is normally found in the ionization region of an electron ionization (EI) or chemical ionization (CI) mass spectrometer. The term encompasses electrospray and atmospheric pressure chemical ionization.

Incorrect: **API** – when used as an abbreviation for atmospheric pressure chemical ionization (APCI). APCI was originally called atmospheric pressure ionization and abbreviated as API. This use is archaic and causes confusion.

Correct: **atmospheric pressure chemical ionization** (**APCI**) – an ionization technique in which analytes are ionized in ion/molecule reactions that take place at atmospheric pressure as opposed to the 0.1–1 Torr pressure used when analytes are introduced into the ion source in the gas phase. In APCI, the analytes are pneumatically sprayed into the heated (~400 °C or greater) ion source. The analyte molecules are volatilized. The nitrogen and oxygen molecules in the ion source are ionized with a corona discharge. The oxygen and nitrogen ions react with solvent molecules, which are now in the gas phase, to form the reagent ions. These reagent ions react with the analyte molecules to produce the analyte ions. APCI allows for liquid-inlet flow rates of 400–2,000 μL min^{-1}. In most cases, only single-charge ions are formed. APCI is a mass-dependent technique (i.e., as long as the amount of analyte does not change, the signal will remain the same regardless of the concentration of the solution). **APcI** is an inappropriate abbreviation for atmospheric pressure chemical ionization. This convention has been adopted by Micromass/Waters and should not be proliferated. This technique was developed by a group under Evan Horning at Baylor University, Houston, Texas, in the mid-1970s (Horning, 1973). The technique was first commercialized by SCIEX; however, it did not gain popularity as an LC/MS technique until the development of electrospray.

Correct: **electrospray** (**ES**) – a process by which ions that are in solution are caused to be in the gas phase (for subsequent *m/z* analysis) through either a mechanism of ion desorption or ion evaporation. The process is carried out by the production of a fine spay of the solution that contains the analyte from a narrow-bore needle (0.1–0.3 mm dia.) to which a potential of 3–5 kV has been applied (6–8 kV in sector instruments). A counter electrode is used to create a field to facilitate the spray formation. The charged droplets evaporate to a point where the number of electrostatic charges on the surface become so large relative to the droplet size that an explosion occurs to produce a number of smaller droplets that also have a surface that contains electrostatic charges. This process repeats until the analyte ion escapes the droplet (ion desorption) or all the solvent has evaporated to leave the ion in the gas phase (ion evaporation). Although the abbreviation **ESI** has been used for **electrospray interface** and **electrospray ionization**, it is believed that the process is not really involved with the formation of ions—just ion desolvation. However, if ionization is any method giving gas-phase ions, the term should be **electrospray ionization**. The first time the abbreviation **ESI** is used, it should be defined. The typical flow rate for electrospray is about 1–10 µL min^{-1}. When flow rates of <1 µL min^{-1} are involved (such as in the case of capillary electrophoreses), a make-up solution of the same composition as the mobile phase is introduced through the annular space formed by placing the ES needle in another needle to produce a sheath flow. In ES, ions with multiple charges can be formed. Because a mass spectrometer measures the abundance of ions based on their mass to number of charges ratio, ions of very high mass can be detected with conventional *m/z* analyzers (i.e., an ion with 10 positive charges and a mass of 5,000 Da would be observed at an *m/z* value of 500 in a mass spectrometer). ES is a concentration-dependent technique (i.e., as the concentration of the analyte in solution decreases, even if the same amount is present, the signal decreases). Although there were several events that contributed to electrospray, its development is credited to John Fenn (Whitehouse, 1984). The technique gained popularity after Fenn reported the detection of multiple-charge ions in 1988 (Meng, 1988).

*Incorrect: **spray ionization** – when used as a general term to describe ionization that results from the spraying of a solution into a mass spectrometer; a generic term for APCI, ES, and thermospray. The "MS Terms and Definitions" that appears on the ASMS Web site (http://www.asms.org) under Items of Interest defines spray ionization as: "method used to ionize liquid samples directly by electrical, thermal, or pneumatic energy through the formation of a spray of fine droplets." This term is too encompassing and can result in some degree of confusion.

Incorrect: **IonSpray™ (ISP)** – when used as a synonym for pneumatically assisted electrospray. IonSpray is the trade name used by PE/SCIEX to describe pneumatically assisted electrospray. The pneumatically assisted electrospray technique allows for liquid flow rates of 10–1,000 µL min^{-1}. IonSpray was coined and developed by Jack Henion at Cornell University, Ithaca, New York (Bruins, 1987). Today, all commercially available ES mass spectrometers have a pneumatic assist. IonSpray should only be used to describe the trademarked product for PE/SCIEX.

Correct: **microelectrospray (microES)** – the low-flow electrospray technique that does not use a make-up solvent where a pump is used to establish the sample flow. Microelectrospray is the first low-flow-rate ES technique described (Emmett, 1994; 1997). The reported flow rates are 300–800 nL min^{-1}.

Correct: **nanoelectrospray (nanoES)** – the low-flow electrospray technique that does not use a make-up solvent where the sample flow is dependent on the potential on the tip of the electrospray needle (Emmett, 1997). The nanoelectrospray name for this technique was originally coined by the developers, Matthias Mann and Matthias Wilm (Wilm, 1996), at the European Molecular Biology Laboratory (EMBL) in Heidelberg, Germany. This technique differs from microES in the diameter of the needle (~1–3 µm) and the flow rates (~25 nL min^{-1}). **NanoSpray™** is the Bruker DALTONICS trade name for nanoelectrospray. This electrospray technique has also been referred to as **MicroIonSpray™** and **NanoFlow™ ES** (trade names that refer to the technique as marketed by PE/SCIEX and Micromass, respectively). These three trade names should only be used to describe the trademarked products.

Note: Before using a term to describe an **electrospray** or a **low-flow electrospray** technique, care must be taken to make sure that the actual process is clearly understood, and the process is not supposed to be implicit by the name.

Correct: **charge-reduction electrospray mass spectrometry (CREMS)** – a technique used to reduce the number of charge sites on a multiple-charge ion produced by electrospray. Although this process has been reported by at least one other investigator (Stephenson, 1997), the term was coined in relation to the use of a polonium-210 particle source (Scalf, 1999). The technique makes the determination of mass as a function of charge site easier while taking advantage of the electrospray process.

Correct: **direct-liquid-inlet (DLI) probe** – a device to introduce a liquid into the ion source of a mass spectrometer. This probe is most often used as a liquid chromatography interface for an electron ionization (EI) or chemical ionization (CI) source.

Correct: **particle beam (PB) interface** – a technique used in LC/MS by which analyte molecules are desolvated in a heated chamber and passed through a momentum separator to remove the solvent vapor from clumps of moist molecules. These clumps of moist molecules enter a conventional EI or CI source, where they are volatilized by impacting onto a heated splatter plate. The PB interface used with EI is the only LC/MS technique that produces a conventional EI mass spectrum. Waters Corporation uses the trademarked term **ThermaBeam**™ to describe this interface technique. **MAGIC (monodisperse aerosol generation interface for combined LC/MS)** was the name applied to this technique (Willoughby, 1984) and was used commercially by Hewlett-Packard (now known as Agilent Technologies).

Correct: **thermospray (TSP)** – an LC/MS technique operated in a "filament-on" or "filament-off" mode. In the **filament-on** mode, analyte ions are formed in a reduced pressure environment through the direct ionization of the mobile phase with an electron beam or corona discharge followed by ion/molecule reactions between the nascent reagent ions and the analyte molecules in the gas phase. In the **filament-off** mode of operation, thermospray produces ions in the gas phase through a process similar to ES (i.e., ions in solution are caused to enter the gas phase for subsequent *m/z* analyses). The ion evaporation or desolvation in thermospray is accomplished at a reduced pressure and without the aid of a high potential. This technique is no longer commercially available. Thermospray was developed by Marvin Vestal at the University of Houston, Houston, Texas (Blakely, 1983).

Correct: **chemical ionization (CI)** or **positive-ion CI (PCI)** – ionization of an analyte that occurs as a result of an ion/molecule reaction. A reagent gas is ionized by an electron beam. The resulting ions react with neutral molecules of the reagent gas to form reagent ions (e.g., CH_5^+ from $CH_4^{+\bullet} + CH_4$). These reagent ions react with analyte molecules to produce analyte ions. The typical CI source pressure in beam-type instruments is 0.1–1 Torr. The reagent gas partial pressure in an internal ionization ion-trap mass spectrometer is $\sim 10^{-5}$ Torr. Chemical ionization was developed by Burnaby Munson and Frank Field at the University of Texas, Austin, Texas (Munson, 1966). There are four different processes that produce positive ions from chemical ionization:

charge transfer (The reagent ion, missing an e^-, takes an e^- from the analyte.)

$$CH_4^{+\bullet} \quad + \quad RH \longrightarrow RH^{+\bullet} \quad + \quad CH_4 \qquad M^{+\bullet}$$

proton transfer (Common when the analyte molecule has a higher proton affinity (PA) than the reagent gas. The analyte takes H^+ from the reagent ion.)

$$CH_5^+ \quad + \quad RH \longrightarrow RH_2^+ \quad + \quad CH_4 \qquad (M+1)^+$$
$$C_2H_5^+ \quad + \quad RH \longrightarrow RH_2^+ \quad + \quad C_2H_4$$

hydride abstraction (The reagent ion has a high hydride affinity, which is the ability to remove a H^- from the analyte molecule.)

$$CF_3^+ \quad + \quad RH \longrightarrow R^+ \quad + \quad CF_3H \qquad (M-1)^+$$
$$C_2H_5^+ \quad + \quad RH \longrightarrow R^+ \quad + \quad C_2H_6$$

collision-stablized complexes (Occurs when the PA of the analyte and reagent gas are comparable. The reagent ion becomes attached to the analyte. When methane is used, the $(M+1)^+$, $(M+29)^+$, and $(M+41)^+$ series is a very good confirmation of the nominal mass of the molecule.)

$$C_2H_5^+ \quad + \quad RH \longrightarrow [C_2H_5{:}RH]^+ \qquad\qquad (M+29)^+$$
$$C_3H_5^+ \quad + \quad RH \longrightarrow [C_3H_5{:}RH]^+ \qquad\qquad (M+41)^+$$

Correct: **negative-ion CI** (**NCI**) – refers to the formation of negative ions by the reaction between an analyte molecule and an anion that was formed from a reagent gas.

negative ion/molecule reaction
$$AB \quad + \quad C^- \longrightarrow ABC^- \text{ or } [AB-H]^- + HC$$

Incorrect: **negative-ion CI** (**NCI**) – when used to describe the formation of negative ions by the analyte molecule's capture of a low-energy electron. The reason this process has been called negative-ion CI is that the low-energy electrons can be produced by the ionization of a gas (usually methane) under high-pressure conditions similar to those used in conventional CI when methane is used as the reagent gas.

Correct: **electron capture/negative-ion** (**ECN**) **detection**, **electron capture/ negative ionization** (**EC/NI**), or **resonance electron capture ionization** (**RECI**) – refers to producing negative ions by the reaction of low-energy electrons with molecules in a mass spectrometer. These terms are the correct terms to describe the formation of negative ions when an analyte molecule captures a low-energy electron. This process, first reported as a GC/MS technique by Ralph Dougherty (Dougherty, 1972) and later by Don Hunt and George Stafford (Hunt, 1976), usually happens under high-pressure conditions that are similar to those used in conventional CI. The resulting negative molecular ions are symbolized as $M^{-\bullet}$, which indicates an odd-electron negative ion. The results of electron capture ionization are:

resonance electron capture
$$AB \quad + \quad e^- \; (\sim 0.1 \text{ eV}) \longrightarrow AB^{-\bullet}$$

dissociative electron capture
$$AB \quad + \quad e^- \; (0\text{–}15 \text{ eV}) \longrightarrow A^\bullet + B^-$$

ion-pair formation that results from electron capture
$$AB \quad + \quad e^- \; (>10 \text{ eV}) \longrightarrow A^- + B^+ + e^-$$

Correct: **chemical reaction interface mass spectrometry (CRIMS)** – a process in which the eluate from a chromatographic process (LC or GC), which has been enriched with respect to the analyte, is passed into a microwave reaction chamber that contains a reaction gas selected to produce a simple product gas such as CO_2, NO, SCl, etc. from organic analytes and their metabolites that contain stable isotopically labeled analytes. A measure of the ratio of naturally occurring product gas to isotopically labeled product gas indicates the presence of metabolites (Abramson, 1994).

Correct: **desorption chemical ionization (DCI)** – a technique by which a nonvolatile thermally labile analyte produces a CI mass spectrum. The analyte is dissolved in a volatile solvent. A drop of the resulting solution is put on a highly conductive loosely coiled element, and the solvent is evaporated. The element, now containing only the analyte, is introduced into a conventional ion source of a mass spectrometer via an insertion probe. In the presence of CI reagent ions at 0.1–1 Torr, a current is passed through the element. Analyte ions produced from ion/molecule reactions are desorbed from the surface of the element into the gas phase for *m/z* analysis. This technique has produced mass spectra of sugars without the need for derivatization.

Correct: **desorption/ionization (DI)** – describes the ionization of an analyte in a solid matrix or solution followed by the subsequent desorption of the ions into the gas phase for *m/z* separation and detection.

Correct: **direct infusion** – a sample-introduction technique. It is the introduction of a liquid sample into the area of the mass spectrometer, where ions are brought into (or produced in) the gas phase. Direct infusion results in a continuous concentration of analyte in the ion source/interface of the mass spectrometer.

Correct: **direct-insertion probe** – one of two types of probes used to introduce solids into the EI or conventional CI ion source of a mass spectrometer. A sample is placed in a small glass tube (similar to a melting-point capillary tube) that is about 10 mm in length. The tube is then fitted to the end of a probe and is inserted into an interface mounted on the ion-source housing of the mass spectrometer. The tip is evacuated, a vacuum interlock is opened, and the probe is pushed into the ion source to position the tube containing the sample near the electron beam. The tip of the probe is then heated to volatilize the sample. This type of probe can easily contaminate the mass spectrometer because of the volume of sample delivered to the ion source. This probe is sometimes referred to as a **solids probe**.

Correct: **direct-exposure probe** – the other type of probe used to introduce solids into the EI or conventional CI ion source. A sample is dissolved in a solvent. A drop of the solution is placed on the end of the probe (usually a rounded glass tip). The solvent is evaporated leaving a thin film of uniform thickness on the inside of the probe. The probe is inserted into the ion-source housing using the same mechanisms as the direct-insertion probe. The tip is then heated to volatilize the analyte. This technique is preferred for the introduction of solids because it will produce far less contamination. A variation of this probe is used with desorption chemical ionization (DCI). The rounded glass tip is replaced with a filament. The sample solution is then put on the filament, the probe is inserted into the ion-source housing, and a current is passed through the filament to aid in the production of gas-phase ions (see **desorption chemical ionization**, above).

Correct: **electron ionization** (**EI**) – ionization of analyte molecules in the gas phase (10^{-3}–10^{-6} Torr) by electrons accelerated between 50 and 100 V. The original standard was 70 V. After the development of chemical ionization as a mass spectrometry technique (1966), papers appeared that used **electron ionization** as opposed to **chemical ionization**. Previously, **EI** was used as an abbreviation for **electron impact**. The use of electron impact is the reason why "EI ionization" is seen in print. The use of **electron impact** or **EI as an abbreviation for electron impact** is incorrect. The *Current IUPAC Recommendations* states, "Electrons and photons do not 'impact' molecules or atoms. They interact with them in ways that result in various electronic excitations, including ionization." Although many refinements were made that led to the current EI source designs, the original development is attributed to Arthur Jeffrey Dempster (1886–1950), who developed the electron bombardment source at the University of Chicago, Chicago, Illinois (Dempster, 1922).

Correct: **field desorption** (**FD**) – an ionization technique by which analyte ions are desorbed from the surface of one of the two electrodes used to produce an electrical field of 10^7–10^8 V cm^{-1}. **FD** is the first desorption technique to be seriously considered in mass spectrometry. FD uses specially prepared sample emitters that allow for the production of protonated molecules (MH$^+$). FD has been used for the analysis of thermally labile nonvolatile samples. This technique has been largely replaced by **ES**. Formation of ions via an electrical field had its origin with field-ion microscopy developed in 1951 by E. W. Müller in Germany. The first developments involving FD were by R. Gomer and M. G. Inghram (Gomer, 1954; 1955; Beckey, 1977).

Correct: **field ionization** (**FI**) – an ionization technique that results in a high abundance of molecular ions. Ionization of an analyte molecule in the vapor phase takes place in an electrical field (10^7–10^8 V cm^{-1}) maintained between two sharp points or edges of two electrodes. The technique of FI was developed by H. D. Beckey, Institut für Physikalische Chemie der Universität Bonn, Bonn, Germany, in 1957 (Beckey, 1963; 1977).

Correct: **flow injection** – a process by which a sample is injected into a continuous liquid flow that enters into the mass spectrometer. Sample introduction is the same as a liquid chromatograph injection; however, in flow injection, there is no column between the injector and the mass spectrometer.

Incorrect: **flow injection** – when used to refer to the technique of direct infusion.

Correct: **fast atom bombardment** (**FAB**) – **FAB mass spectrometry** – a desorption/ionization technique that is used to obtain ions of large, nonvolatile, thermally labile analytes in the gas phase for *m/z* analyses. A solution of the analyte is mixed with a matrix (usually glycerol) and placed in an evacuated ion-source housing. The surface of this mixture is bombarded with a stream of atoms (usually argon) that have been given approximately 5–8 kV of translational energy. The impact of the particles on the surface produces molecular sputtering, which results in desorption of ions from the liquid/vacuum interface into the gas phase. The analyte may be in ionic form in the target sample, or it is ionized by proton transfer that accompanies the bombardment. A related technique, **liquid secondary-ion mass spectrometry** (**LSIMS**), employs a stream of ions (often cesium) to accomplish the ionization. FAB and LSIMS have largely been replaced by **ES** and **MALDI**. The FAB technique is attributed to Mickey Barber, University of Manchester Institute of Science, Manchester, England (Barber, 1981).

Correct: **continuous-flow FAB** (**CF-FAB**) – largely an LC/MS technique (but could be used to improve signal-to-background in infusion experiments) in which the eluate from the liquid chromatograph is continually mixed with the FAB matrix material (usually glycerol), and the mixture passes through a needle with a slanted tip. The bombardment is pulsed to produce each mass spectrum as a continually changing sample flows to the FAB target. This technique has been largely replaced by **ES**. The development of CF-FAB is jointly attributed to Richard Caprioli at the University of Texas Medical School, Houston, Texas (Caprioli, 1986); and M. Ito, who developed Frit-FAB in Japan (Ito, 1985).

Correct: **matrix-assisted laser desorption/ionization** (**MALDI**) – a desorption/ionization technique in which a laser is used to produce ions from analytes that are present in a solid matrix. Analyte molecules undergo electronic excitation with a UV laser or vibrational excitation with an IR laser—clearly a process yet to be understood. The matrix acts as an energy disperser to bring about the ionization of the analyte. The laser provides energy for the ionization and the desorption of the ions from the matrix. This technique produces predominantly single-charge ions; however, some double- and triple-charge ions may be observed. MALDI was developed by Franz Hillenkamp and Michael Karas at the Institute für medizinische Physik und Biophysik, Universität Münster, Münster, Germany (Karas, 1987; Hillenkamp, 1991).

Correct: **plasma desorption** (**PD**) – a technique that uses the fission products of ^{252}Cf to produce ions of large molecular weight nonvolatile analytes. PD uses a TOF-MS as the *m/z* analyzer. An average spectrum results from several hours of data acquisition. The technique was developed by Robert Macfarlane at Texas A&M University, College Station, Texas, in the mid-1970s (Sundqvist, 1985). The technique is no longer in use and was replaced by ES and MALDI.

*Correct:** **collisionally activated dissociation** (**CAD**) and **collision-induced dissociation** (**CID**) – terms that describe ion fragmentation in an MS/MS experiment. The precursor ion has its translational energy converted to internal energy by collisions with neutral molecules to bring about a dissociation. CAD is used interchangeably with CID. An ion can undergo collisional activation or collisional excitation without fragmenting. With two terms used to describe the same event, confusion can result. Because ions can be collisionally activated or be in a state of collisional excitation, **collisionally activated dissociation** is the preference of this author. The two terms are considered equal in the *Current IUPAC Recommendations*.

Correct: **ion/molecule reaction** – describes an ion/neutral reaction where the neutral species is a molecule.

Incorrect: **ion-molecule reaction** – when used to suggest a reaction of a species that is both an ion and a molecule because of the presence of the hyphen. Therefore, the term **ion-molecule reaction** describes a different event than **ion/molecule reaction**.

Correct: **mass analysis** – a process by which a mixture of ionic or neutral species is identified according to the mass-to-charge ratios (ions) or their aggregate atomic masses (neutrals). The analysis may be quantitative and/or qualitative.

Correct: **mass spectrometry** – the study of matter based on the mass of molecules and on the mass of the pieces of the molecule. Mass spectrometry is often involved with mass spectra obtained with a mass spectrometer. The *Current IUPAC Recommendations'* definition of mass spectroscopy could more appropriately be applied to mass spectrometry than its definition of mass spectrometry.

Incorrect: **mass spectroscopy** – when used to imply the use of an optical device. This conclusion is reached based on the definition of spectroscopy found in any dictionary. There are no light sources in mass spectrometry. Photoionization has been reported; however, it is not an often-used means of producing ions in a mass spectrometer ($M + h\nu \rightarrow M^{+\bullet} + e$). Mass spectroscopy has been used in a loose sense to include the use of mass spectrometers (abundance positive-ray analysis) and mass spectrographs (accurate-mass positive-ray analysis) as well as the studies of isotopic abundance, precise mass determination, analytical chemical use, appearance potential, etc. Mass spectroscopy is too encompassing for general use. Although declining in use, mass spectroscopy is still a popular term, especially with a large segment of the European mass spectrometry community. The *Current IUPAC Recommendations* (see following sidebar) refers to the term **mass spectroscope** as essentially obsolete, but uses it in the definition of mass spectrometry and then goes on to imply that the term **mass spectroscopy** is appropriate.

Correct: **mass spectrometer** – a term first used by two well-known early mass spectrometry pioneers, William R. Smythe (U.S. scientist) and Josef Heinvich Elizabeth Mattauch (Austrian physicist) ca. 1926 (Kiser, 1965), and is applied to those instruments that bring a focused beam of ions to a fixed collector, where the ion current is detected electrically. The term **mass spectrometer** is now used to describe all instruments that measure the abundance of ions based on their *m/z* values.

Incorrect: **mass spectrograph** – when used to describe a specific type of *m/z* analyzer and should not be applied to a mass spectrometer. A mass spectrograph (first introduced by F. W. Aston in 1921) is an instrument that produces a focused mass [*m/z*] spectrum on a focal plane, where a photographic plate may be located. These instruments are capable of a high degree of mass accuracy, but are not very good at determining ion abundances.

> **From the Latest IUPAC Recommendations**
>
> (considered to be obsolete and/or misleading by this author)
>
> *Mass spectrometry.* "The branch of science that deals with all aspects of mass spectroscopes and the results that are obtained with these instruments." *This definition is somewhat contradictory with respect to the following definition.*
>
> *Mass spectroscope.* "A term, which is now essentially obsolete, that refers to either a mass spectrometer or a mass spectrograph."
>
> *Mass spectroscopy.* "The study of systems that cause the formation of gaseous ions, which are characterized by their mass-to-charge ratios and relative abundances, with or without fragmentation." *This definition is somewhat contradictory to that of mass spectroscope.*
>
> *Mass spectrograph* and *mass spectrometer* are defined above.
>
> From the ambiguity of the above definitions, along with their contradictions, you have to wonder what the authors really mean.

Correct: *m/z* **analyzer** or *m/z* **analysis** – terms that better describe the mass spectrometer and its functions than the terms **mass analyzer** or **mass analysis**. The use of *m/z* **analyzer** and *m/z* **analysis** is supported through the continual proliferation of mass spectrometers that use the electrospray technique that produce multiple-charge ions of high mass.

Incorrect: **mass spectrophotometer** – a term that never had any official recognition in mass spectrometry. There are no light bulbs in a mass spectrometer.

Correct: **mass spectrometrist** – the term used to describe a person who uses mass spectrometry. The term **mass spectroscopist** is incorrect.

Correct: **MS/MS (mass spectrometry/mass spectrometry)** – a mass spectral technique in which ions are caused to change mass (usually via decomposition, but can form a heavier product from collision with a "reactive" neutral) to produce information that may not be obtainable from an initial ionization of the analyte. MS/MS is a technique where ion formation/fragmentation and subsequent decomposition of the original ions is carried out "in tandem." Using multiple *m/z* analyzers and an ion beam results in "tandem in space" (e.g., triple-quadrupole and quadrupole-TOF mass spectrometers). Using ion-trap mass spectrometers (quadrupole or magnetic) results in "tandem in time."

There are three types of **MS/MS** analyses:

> **product-ion analysis**: This technique involves the isolation of a precursor ion, the collisionally activated dissociation of the precursor ion, and the production of a mass spectrum of the ions that result from this induced fragmentation. This type of analysis can be conducted in an instrument that involves the transportation of an ion beam from an area where the precursor ion is selected into the collision region followed by an area where the product ions are sorted, such as a: 1) triple-quadrupole mass spectrometer,

2) reverse-geometry double-focusing mass spectrometer, or 3) hybrid instrument composed of a quadrupole and time-of-flight mass spectrometer. All of these instruments use two *m/z* analyzers separated by a collision cell. This analysis can also be carried out in a quadrupole ion-trap or Fourier transform ion-cyclotron resonance (FTICR) mass spectrometer. When carried out in one of these two latter types of instruments, a specific product ion is isolated by ejecting all other ions and a dissociation is carried out. The dissociation process can be repeated with one of the product ions. The process can be carried out on several more generations of product ions. This process of generation of ions from the fragmentation of various generations of product ions is called **MS^n** where **n** is the number of ion formations.

precursor-ion analysis: All of the ions formed by a primary ionization pass, one *m/z* value at a time, into the collision cell. The second *m/z* analyzer is set to allow only a single *m/z* value to reach the detector. This process results in the identification of only precursor ions that produce product ions of a specific *m/z* value. This type of MS/MS analysis can only be performed in instruments that use an ion beam.

common-neutral-loss analysis: All of the precursor ions formed by a primary ionization are allowed to pass, one *m/z* value at a time, into the collision cell. The second *m/z* analyzer is set to allow all ions to pass to the detector, one *m/z* value at a time. However, the *m/z* value of the second analyzer is offset by a fixed *m/z* difference from that of the first analyzer. This process results in the detection of analytes that have a common neutral loss. This type of MS/MS analysis can only be performed in instruments that use an ion beam.

MS/MS can also be carried out in a hybrid instrument that uses an ion trap in conjunction with a beam-type *m/z* analyzer, thereby allowing for all three types of MS/MS analyses.

The symbolisms proposed by de Hoffmann (de Hoffmann, 1996) for the three different types of MS/MS analyses and that of R. W. Kondrat and R. G. Cooks (Kondrat, Cooks, 1978) are shown in **Figures 3A** and **3B**, respectively.

Correct: **in-source CAD** – similar to **MS/MS** in that the results are product ions produced by a collision of precursor ions. **In-source CAD** is used with the **API** techniques of **APCI** and **ES**. There is no *m/z* selection of a precursor ion by mass spectrometry. The only selection is accomplished through chromatographic purification of the precursor molecule. The precursor ions are fragmented in the moderate pressure area of the ion source because they undergo activating collisions with the gas (in most cases, nitrogen molecules) that is present. The fragmentation efficiency is based on a controllable velocity of the ions as they enter the *m/z* analyzer. This technique requires a limit to the number of different *m/z* values of precursor ions.

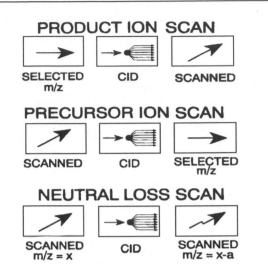

PRODUCT ION SCAN

SELECTED m/z — CID — SCANNED

PRECURSOR ION SCAN

SCANNED — CID — SELECTED m/z

NEUTRAL LOSS SCAN

SCANNED m/z = x — CID — SCANNED m/z = x-a

● → ○ PRODUCT ION SCAN
○ → ● PRECURSOR ION SCAN
○ → ○ NEUTRAL LOSS SCAN
● → ● SELECTED REACTION MONITORING

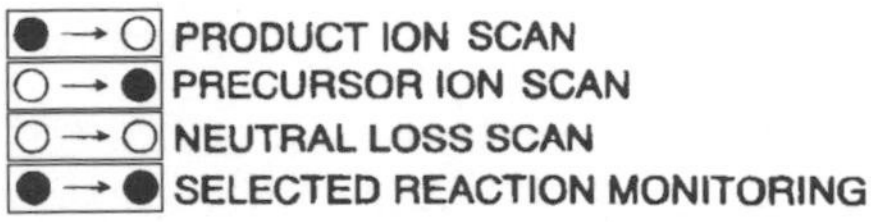

Figure 3B. Symbolism proposed by Kondrat and Cooks to represent the various scan modes; ● refers to a fixed mass analyzer, and ○ refers to a scanning mass spectrometer. (Kondrat, RW; Cooks, RG *Anal. Chem.* **1978**, *50*, 81A)

Figure 3A. MS/MS symbolism proposed by E. de Hoffmann. (de Hoffmann, E "Tandem Mass Spectrometry: A Primer," *J. Mass Spectrom.* **1996**, *31*, 129–137)

Correct: **surface-induced dissociation** (**SID**) – the fragmentation of a precursor ion brought about by a collision with a solid surface. SID has the advantage of not requiring a relatively high-pressure collision gas, and the efficiency approaches 100% when large energy transfers occur.

Correct: **selected reaction monitoring** (**SRM**) – an **MS/MS** technique that is similar to selected ion monitoring (SIM), but may allow for a much higher degree of specificity. In SRM, ions of a specific *m/z* value are allowed to pass into the collision cell. The second *m/z* analyzer is set to allow only product ions of a specific *m/z* value to pass to the detector (i.e., the precursor ion must undergo a selected transformation or reaction to result in a response from the detector). Regardless of whether there is more than one precursor ion and multiple product ions from a single precursor ion, the term **SRM** applies, just as the term **SIM** applies, when there is more than one ion monitored.

Incorrect: **multiple reaction monitoring** (**MRM**) – when used as a synonym for **SRM**. This neologism was created by an instrument manufacturer to distinguish their instruments from instruments of other manufacturers. This term is misleading as to what is actually being monitored in the mass spectrometer and should not be used. In the case where the **SRM** analysis involves multiple generations of product ions (e.g., ion-trap mass spectrometers), the term **MRM** has been used to show that more than one generation of product ions are being monitored, which gives the term **MRM** two different meanings and leads to even more confusion.

Note: In **MS/MS** analyses, ions collide with neutral molecules in a collision cell; and in **in-source CAD**, ions collide with neutral molecules in the ion source. When these collisions take place, a number of different things can happen to the precursor ion. The most common is collisionally activated dissociation.

$$m_p^+ \; + \; N \longrightarrow m_d^+ \; + \; m_n \; + \; N$$

However, the following three reactions can also take place. Because of the nature of API techniques, these reactions can be significant in in-source CAD.

charge exchange:

$$m_p^{+\bullet} \; + \; N \longrightarrow m_p \; + \; N^{+\bullet}$$

partial charge transfer:

$$m_p^{2+\bullet} \; + \; N \longrightarrow m_p^{+\bullet} \; + \; N^{+\bullet}$$

charge stripping: This process is an ion/molecule reaction that increases the number of positive charges on an ion. Charge stripping is an ionization process.

$$m_p^{-\bullet} \; + \; N \longrightarrow m_p^{+\bullet} \; + \; N \; + \; 2e^-$$

$$m_p^{+\bullet} \; + \; N \longrightarrow m_p^{2+\bullet} \; + \; N \; + \; e^-$$

$$m_p \; + \; N \longrightarrow m_p^{+\bullet} \; + \; N \; + \; e^-$$

It is possible that charge stripping can be combined with a mass change.

$$m_p^- \; + \; N \longrightarrow m_d^+ \; + \; (m_p - m_d) \; + \; N \; + \; 2e^-$$

If the collision gas is a chemically active substance, then an association reaction can take place to yield an adduct ion (or even a condensation product) that has a mass that is greater than the precursor ion.

$$m_p^+ \; + \; m_n \longrightarrow (m_p^+ + m_n)$$

Another ion/neutral reaction is **charge inversion**. This process is where the sign of the charge is reversed. The first example of charge stripping, above, is also charge inversion.

All of these ion/neutral reactions are **charge permutations**. This general term describes an ion/neutral species reaction where a change occurs in the magnitude and/or sign of the charge.

Correct: **peak matching** – a technique used with double-focusing instruments operating in the high-resolution mode (typically $m/\Delta m \geq 10000$) where a particular monoisotope peak of an unknown is superimposed to a peak of an internal standard (the exact mass of which can be calculated); the two molecules, unknown and internal standard, must be different but close in mass. The magnet position is fixed, and the accelerating voltage is switched between a known and reference peak. The accelerating voltage difference can be used to calculate the mass difference between the unknown peak and the standard peak. The peak matching must first be calibrated using two reference points to define the slope and intercept, which is done prior to making a series of measurements. A second magnet coil is sometimes used, but its purpose is just to sweep the peak over a narrow range so that you can see the peak shape on the oscilloscope. The mass measurement comes from the accelerating voltage (and electric sector) switching.

Correct: **secondary-ion mass spectrometry (SIMS)** – a technique that uses the bombardment of a solid sample with a high-energy beam of ions to produce a mass spectrum of secondary ions that are generated on the surface of the sample. This technique was originally developed for the analysis of inorganic materials and polymeric surfaces. **Static SIMS** uses lower primary-ion-dose densities at pressures of 10^{-10} Torr. This technique is employed for surface analysis in the *x-y* plane and is used in organic applications. **Dynamic SIMS** uses a higher ion flux to achieve depth-profile measurements along the *z* axis of the sample.

Correct: **selected ion monitoring (SIM)** – a chromatographic/mass spectrometric technique where ion current at only one or a few selected *m/z* values is detected and stored during the chromatographic separation. Either the full phrase or the abbreviation can be used in print. When spoken, each letter should be pronounced, and the acronym should not be used. It is easy to confuse **SIM** with **SIMS**, which is spoken as the acronym for **secondary-ion mass spectrometry**.

Incorrect: **selective ion monitoring** – when used as a phrase that indicates the ions are doing the selecting of what to monitor, not that they were selected by the analyst for observation. This phrase has appeared occasionally in print but is incorrect. The term **mass-selective detector (MSD)** is correct because it refers to the fact that the detector monitors only ion currents at certain "masses" [*m/z* values where *z* is always 1].

Note: **MID (multiple-ion detection)** – a product, originally sold by LKB Instruments, that performed an SIM analysis. Finnigan uses this term to describe the monitoring of the ion current of a few specific *m/z* values. The term **MID** is considered to be archaic and should not be used.

Another term used to describe the SIM technique is **mass fragmentography**. Mass fragmentography is considered to be archaic and should not be used. A third term used to describe the technique of selected ion monitoring in magnetic-sector instruments is **accelerating voltage alternation (AVA)**. AVA is highly specific to the instrument type and is no longer considered to be correct.

SIM has also been used as an abbreviation for the term **single-ion monitoring**. Single-ion monitoring is too restrictive and should not be used.

Correct: **octet rule** – the statement that no energy shell of an atom can hold more than eight electrons, as long as it is the outer most shell of the atom.

Correct: **homolytic** (a.k.a. **radical-site-driven**) **cleavage** – a fragmentation that results from one of a pair of electrons between two atoms moving to form a pair with the odd electron. After fragmentation, the atom that contains the charge when the ion is formed retains the charge. A radical is lost as a result of the fragmentation. This reaction involves the movement of a single electron and is symbolized by a single-barbed arrow—the so-called "fishhook" convention (Budzikiewicz, 1964).

Unfortunately, the *Current IUPAC Recommendations* states that the symbol that indicates the movement of one electron (homolysis) and the symbol for the movement of two electrons (heterolysis) is the same—a double-barbed arrow.

Correct: **alpha** (α) **cleavage** (**a special form of homolytic cleavage**) – a fragmentation (homolytic cleavage) that results from one of the pair of electrons between the atom attached to the atom with the odd electron and an adjacent atom that pairs with the odd electron. After fragmentation, the atom that contains the charge when the ion is formed retains the charge. A radical is lost as a result of the fragmentation. This fragmentation is homolytic cleavage because it involves the movement of a single electron (McLafferty, 1973).

$$R_1 - C \underset{\overset{\|}{\overset{+\cdot}{O}}}{\overset{a}{\underset{b}{\overset{c}{-}}}} CH_2 - R_2 \longrightarrow \begin{array}{c} R_1 - C \equiv \overset{+}{O} : \\ + \\ \cdot CH_2 - R_2 \end{array}$$

Incorrect: **alpha** (α) **cleavage** – when defined as "…fission of a bond originating at an atom that is adjacent to the one assumed to bear the charge; the definition of β, γ, [etc.] then follows automatically" (Budzikiewicz, 1964). This definition allows the use of α cleavage to describe a bond fission that results in original charge-site retention (homolytic cleavage) or charge-site migration as a result of bond fission (heterolytic cleavage). Using α cleavage with this somewhat ambiguous definition can lead to confusion. The convention established by McLafferty (McLafferty, 1973), where the term α **cleavage** is used to define a special case of homolytic fission, results in a clearer communication. In an attempt to reduce the confusion created from the use of this term, some authors use "α cleavage with charge retention" and "α cleavage with charge migration." The *Current IUPAC Recommendations* uses the Budzikiewicz, Djerassi, and Williams recommendation as the definition of α cleavage (Budzikiewicz, 1964).

Correct: **benzylic cleavage** – a fragmentation that takes place at the carbon atom attached to a phenyl group. When the phenyl group is C_6H_5, the benzylic cleavage will result in a benzyl ion with a formula of $C_6H_5=CH_2^+$, which can be isomeric with the tropylium ion. Benzylic cleavage is a special case of homolytic cleavage and is due to the loss of a pi electron, which places the site of the charge and the radical on the phenyl ring.

Correct: **heterolytic** (a.k.a. **charge-site-driven** or **inductive** – **i**) **cleavage** – a fragmentation that results from the pair of electrons between the atom attached to the atom with the charge and an adjacent atom that moves to the site of the charge. This fragmentation involves the movement of the charge site to the adjacent atom. A radical is lost as a result of the fragmentation. The movement of a pair of electrons is symbolized by a double-barbed arrow (Budzikiewicz, 1964).

An example of the fragmentation of a heterolytic cleavage follows.

$$R_1 - \overset{\overset{\displaystyle \cdot\cdot\overset{+\nearrow}{O}}{\|}}{C} \bullet\bullet CH_2 - R_2 \longrightarrow \quad R_1 - \overset{\bullet}{C} = \overset{\cdot\cdot}{\underset{\cdot\cdot}{O}}$$

$$+$$
$$+ CH_2 - R_2$$

OR

$$R_1 - \overset{\overset{\displaystyle \cdot\cdot\overset{+\nearrow}{O}}{|}}{\underset{\cdot}{C}} \bullet\bullet CH_2 - R_2$$

$$\downarrow$$

$$R_1 - \overset{\overset{\displaystyle \cdot\cdot\overset{\cdot}{O}}{|}}{\underset{+}{C}} \bullet\bullet CH_2 - R_2$$

Correct: **γ-hydrogen shift-induced beta (β) cleavage** (a.k.a. the **McLafferty rearrangement**) – a rearrangement reaction that was originally described by an Australian chemist, A. J. C. Nicholson (Nicholson, 1954), but was named after Fred W. McLafferty (McLafferty, 1993) because of the extent to which he studied and reported the reaction in a wide variety of compound types. An odd-electron fragment ion is formed by the loss of a molecule. This fragment results from a γ-hydrogen shift to an unsaturated group such as a carbonyl (when the site of the odd electron and the charge is on the oxygen atom). The γ hydrogen moves the radical site to the carbon atom that originally contained the γ hydrogen. This new location of the radical site initiates an α-cleavage reaction that causes the fragmentation of the carbon-carbon bond that is beta to the unsaturated group and the loss of a terminal olefin.

$$CH_3 - (CH_2)\overline{_{13}} - \overset{\displaystyle CH}{\underset{\displaystyle CH_2}{\overset{\displaystyle \diagup}{\diagdown}}} \overset{\displaystyle H}{\underset{\displaystyle \overset{+}{O}:}{\diagdown}} \quad \xrightarrow{\text{rH}} \quad CH_3 - (CH_2)\overline{_{13}} - \overset{\displaystyle CH}{\underset{\displaystyle \|}{\overset{\displaystyle}{}}} \quad + \quad \overset{\displaystyle H}{\underset{\displaystyle \overset{+}{O}:}{\diagdown}}$$

m/z = 298   m/z = 74

Correct: **EIEIO** – a term associated with the farm or the animals on the farm of a man named MacDonald—affectionately referred to as "Old MacDonald." **EIEIO** has also been used in mass spectrometry as an acronym for **electron-induced excitation in organics** (McLafferty, 1993) and **electron-impact excitation of ions from organics** (Cody, 1979). This technique involves the excitation of trapped ions with a continuous electron beam. The spectra obtained by EIEIO in an ion-cyclotron resonance mass spectrometer are analogous to spectra that result from CAD and yield characteristic structural information. The technique is also known as **electron-induced dissociation**.

Correct: **Field's rule** – associated with the fragmentation of even-electron ions. The tendency for a neutral fragment to leave depends on its proton affinity (PA). The formation of $C_2H_5^+$ is greater from $C_2H_5O^+{=}CH_2$ (through charge migration) than from $C_2H_5S^+{=}CH_2$ because the PA of $O{=}CH_2$ is less than that of $S{=}CH_2$ (7.4 eV vs 8.9 eV). The lower the PA of the neutral molecule, the greater the tendency for it to leave the even-electron ion (Field, 1972).

Correct: **ortho effect** – a term associated with the effect of a group with a labile hydrogen adjacent to a group that can carry the site of the charge in an electron ionization on an aromatic ring. This adjacent relationship causes a hydrogen rearrangement, resulting in the loss of a molecule by charge migration to produce an odd-electron fragment ion. This behavior is not seen when the two groups are para or meta to one another. Examples are seen in the mass spectra of o-chlorophenol and salicylic acid.

Correct: **pi bond** – a bond formed by the side-by-side overlap of the *p* suborbitals of the outer energy shells of two adjacent atoms. A double bond will have one pi bond and one sigma bond; a triple bond will have two pi bonds and one sigma bond.

Correct: **retro-Diels–Alder reaction** – a cleavage that results from the breaking of two bonds to form a butadiene odd-electron ion and a neutral even-electron olefinic fragment (a molecule) or an odd-electron product ion by the explosion of a molecule of a butadiene. This reaction is the reverse of the 1,4-addition of an olefinic unit to a conjugated diene (a Diels–Alder reaction). The retro-Diels–Alder fragmentation is often found in the mass spectra of cyclic olefins (Biemann, 1962).

Correct: **sigma bond** – a bond formed by the end-to-end overlap of the *sp* hybridized suborbitals of the outer energy shell of two adjacent atoms or the overlap of an *sp* hybridized suborbital of the outer shell of an atom and the *s* suborbital of the single energy shell of the hydrogen atom. The sigma bond can be formed using *sp*, *sp*^{2}, or *sp*^{3} hybrid orbitals.

Correct: **sigma-bond cleavage** – a fragmentation that results from the breaking of a sigma bond. It usually occurs in molecular ions when a sigma electron is lost in the ionization process. A radical is lost as a result of the fragmentation.

Correct: **skeletal rearrangements** – rearrangements that involve the shift of a hydrogen atom or a hydride ion that results in a new radical site or charge site, respectively. The McLafferty rearrangement is an example of a skeletal rearrangement, as is the loss of water from a protonated alkyl aldehyde or the periodicity of peaks that differ by 56 *m/z* units in the mass spectrum of methyl stearate. The same symbolism used to indicate the McLafferty rearrangement (shown above) is used to indicate a skeletal rearrangement.

Correct: **Stevenson's rule** – associated with sigma-bond cleavage. The sigma-bond cleavage of an odd-electron ion leads to two sets of ion-radical products: ABCD$^{+\bullet}$ will produce A$^+$ and $^\bullet$BCD, or DCB$^+$ and $^\bullet$A. The radical that has the highest tendency to retain the odd electron will also have the higher ionization potential. Therefore, the ion with the lowest ionization potential will be preferentially formed. This lower energy ion should be more stable; therefore, the more abundant (Stevenson, 1951). A notable exception to Stevenson's rule is the preference for the **loss of the largest alkyl** radical at the site of ionization. In a series of secondary carbenium ions produced from an aliphatic hydrocarbon molecular ion that has a methyl, ethyl, and butyl group along with a hydrogen atom attached to a carbon, the most stable ion would be the one that results from the loss of the hydrogen radical; however, it will be the least abundant ion. The most abundant will be the ion produced by the loss of the butyl radical, which is the largest of the four possible losses. This ion is also the least stable.

Inappropriate: **simple cleavage** – when used to describe the unimolecular fragmentation of single-charge molecular ions (M$^{+\bullet}$) to produce an even-electron fragment ion (EE$^+$) and an odd-electron radical (OE$^\bullet$) by the breaking of a single bond. This term was first introduced by McLafferty (McLafferty, 1973). It is an all-encompassing term that can refer to σ–bond, homolytic, or heterolytic cleavage. With such an overall ambiguous term, its use could result in rampant confusion. Another term (resulting in less confusion) often used in conjunction with simple cleavage is **dissociation with rearrangement**, which is defined as the fragmentation of a M$^{+\bullet}$ through a process in which bonds are broken and new bonds are formed.

IONS

Correct: **acylium ion** – an even-electron ion that is the product of a single-bond cleavage (usually alpha cleavage) of an odd-electron ion that contains oxygen, and the original site of the charge is on the oxygen atom. The acylium ion (usually referred to in EI mass spectrometry) has the form: $R_2-C\equiv O^+$.

Correct: **allyl ion** – an even-electron ion that is the product of alpha cleavage initiated by ionization at an olefinic double bond (or phenyl π-system: **benzyl ion**). Allyl ions have the form: $CH_2=CH-C^+H_2 \leftrightarrow {}^+CH_2-CH= CH_2$.

Correct: **alkyl ion** – an even-electron ion that is the product of a single-bond cleavage (usually sigma-bond cleavage) of hydrocarbon ion (odd- or even-electron) that contains no aromatic groups.

Correct: **appearance energy** – the minimum energy that must be absorbed by an atom or molecule to produce a specified ion. The specified ion does not have to be a fragment ion. The term **appearance potential** is no longer recommended by IUPAC.

Incorrect: **appearance potential** – when used as a term to specify the energy of the electrons in electron ionization at which fragment ions begin to appear. The appearance potential is less than 20 eV for most organic compounds. This term is no longer recommended because of the potential confusion with **appearance energy**.

Correct: **benzyl ion** – an even-electron ion that is the product of alpha cleavage of an odd-electron ion that is initiated by the radical and charge on the ring. The benzyl ion has the form: $C_6H_5-C^+H_2 \leftrightarrow H_2C=C_6^+H_5$. Unlike the tropylium ion (which has all 7 carbon atoms in the ring), the benzyl ion has only 6 carbon atoms in the ring.

Correct: **carbenium ion** – an even-electron hypovalent ion with the charge on a carbon atom usually formed by sigma-bond cleavage. The three types of carbenium ions are shown below where R, R', and R" are generally any organic substructure. The order of stability is tertiary > secondary > primary.

$$R-\overset{+}{C}H_2 \qquad R-\overset{+}{C}H-R' \qquad R-\overset{R''}{\underset{R'}{\overset{|}{\underset{|}{C}}}}+$$

Primary   Secondary   Tertiary

Incorrect: **carbonium ion** – when used to describe a trivalent carbon ion. A **carbonium ion** is an even-electron hypervalent ion formed by the addition at a carbon site of a molecule to form a fifth covalent bond. An example of a carbonium ion is the primary CI reagent ion of methane, CH_5^+ (methanonium ion). For many years, the positive even-electron ion of carbon with three subgroups (which is now known as a **carbenium ion**) was called a carbonium ion. Although the existence of the CH_5^+ ion in the gas phase had first been reported much earlier (Tal'rose, 1952), when George Olah began to describe its existence in organic solution chemistry (Olah, 1971; 1972), he proposed the change to the current nomenclature of carbonium and carbenium based on the evidence of an intermediate in which the positive charge at the carbon is a result of five covalent bonds. At the same time, it was proposed that both carbonium and carbenium ions be referred to as **carbocations**. This nomenclature has been accepted by IUPAC (Gold, 1983). The term **carbonium ion** sometimes (incorrectly) appears in current literature to describe the trivalent **carbenium ion**.

Incorrect: **carbocation** or **carbanion** – when used to refer to an organic positive or negative ion. These terms are incorrect because the use of **cation** as a synonym for a positive ion and the use of **anion** as a synonym for a negative ion is no longer recommended by IUPAC for use in mass spectrometry (see **positive ions** and **negative ions**, below).

Correct: **cluster ion** – an ion formed by the combination of an ion with one or more of another ion, atom, or molecule of a chemical species (e.g., $[(H_2O)_nH]^+$ is a cluster ion).

Correct: **distonic ion** (**distonic radical ion**) – an odd-electron ion in which the radical and charged sites are separated. The site of the charge and the radical site are associated with nonadjacent atoms. Distonic ions result from rearrangements (Radom, 1984).

Distonic Ion

Correct: **electron volt (eV)** – a unit of energy that is the work done on an electron when passing through a potential rise of 1 volt. $1\ eV = 1.602 \times 10^{-19}$ joules. The energy of the electron beam in electron ionization mass spectrometry is expressed in eV. In modern instruments, the ionization energy standard for EI mass spectrometry is 50–100 eV (i.e., electrons are accelerated between 50 and 100 volts in the ion source). In the early days of mass spectrometry, the standard was set at 70 eV.

Correct: **electron energy** – describes the potential difference through which electrons are accelerated in electron ionization. The term **ionization energy** has been used as a synonym for electron energy; however, this use is incorrect because **electron energy** is specific to **electron ionization**, and ionization energy can be applied to any form of ion formation.

Correct: **even-electron ion** (EE^+ or EE^-) – an ion (positive or negative) that contains no unpaired electrons (e.g., CH_3^+ in the ground state). All fragment ions are not necessarily even-electron ions.

Correct: **odd-electron ion** ($OE^{+\bullet}$ or $OE^{-\bullet}$) – an ion (positive or negative) that contains an unpaired electron (e.g., $CH_4^{+\bullet}$). Most single-charge molecular ions (positive or negative) are odd-electron ions. Double-charge molecular ions like $CH_4^{(2+)(2\bullet)}$ are even-electron ions. Because nitric oxide (NO) is a radical, not a molecule, the ion produced by the loss of an electron is not a molecular ion (see **radical ion**, below).

Correct: **fragment ion** (X^+, X^-, $X^{+\bullet}$, or $X^{-\,\bullet}$) – an electrically charged dissociation product of an ionic fragmentation. A fragment ion can dissociate further, can be positive or negative, and can be an even-electron or odd-electron ion. Fragment ions produced from molecular ions are represented by peak intensities at m/z values that correspond to the m/z values of fragments of the analyte molecule that are formed by the loss of a neutral molecule or radical or an ion of the opposite charge.

Correct: **iminium ion** – an even-electron ion that is the product of a single-bond cleavage of an ion that contains nitrogen. The iminium ion (usually referred to in EI mass spectrometry) has the form: $RC\equiv N^+ - R'$.

Correct: **immonium ion** – an even-electron ion that is the product of a single-bond cleavage of an odd-electron ion (EI mass spectrometry) or the protonation or cationization of an analyte molecule (ES, APCI, or CI mass spectrometry) that contains nitrogen. The immonium ion has the form: $R_2C=N^+R'_2$.

Correct: **ion** – a particle that results from an atom or molecule that has an unequal number of positive and negative components. Ions result from the removal or addition of electrons; or the addition of negative ionic species such as Cl^-, etc.; or the addition of protons or other positive ionic species such as Na^+, K^+, Ag^+, etc. to neutral atoms or molecules. Ion is derived from the Greek word that means "go" because charged particles go toward or away from a charged electrode. The term **ion** was first used by Michael Faraday in 1834.

Correct: **isobaric** – describes ions or peaks that have the same integer m/z value but represent different elemental compositions (e.g., carbon monoxide and ethylene single-charge molecular ions both have a nominal m/z value of 28). The terms **isobaric ions** and **isobaric peaks** are correct. Isobaric peaks are peaks with the same integer m/z value that represent two different ions. These ions could have different elemental compositions or different isomers (structural or optical) of the same elemental composition (acetone and propanal).

Correct: **isotope cluster** – a group of peaks close to one another that represent ions with the same elemental composition but a different isotopic composition. In mass spectra of substances containing only C, H, N, O, S, Si, P, and halogens, the lowest m/z value peak (not the most intense peak) is the monoisotope peak (e.g., peaks representing the molecular ions $^{12}CH_3^{35}Cl^{+\bullet}$, $^{13}CH_3^{35}Cl^{+\bullet}$, $^{12}CH_3^{37}Cl^{+\bullet}$, and $^{13}CH_3^{37}Cl^{+\bullet}$ constitute an isotope cluster).

Correct: **isotopic ion** – any ion that contains one or more of the less abundant naturally occurring stable isotopes of the elements that make up the structure of the ion (e.g., $^{13}CCH_5^+$).

Correct: **isotopic molecular ion** – a molecular ion that contains one or more of the less abundant naturally occurring stable isotopes of the atoms that make up the molecular structure of the ion.

Correct: **metastable ion** – an ion that dissociates after leaving the ion source and before reaching the detector (see **Terms Associated with Double-Focusing Mass Spectrometers**).

Correct: *m/z* – the symbol for the mass-to-charge ratio of an ion or peak (*m* = mass of the particle; *z* = number of charges on the particle). Although this term is the official usage as prescribed in the *Current IUPAC Recommendations* and the *ASMS Guidelines*, unfortunately, *m/z* is a mass spectrometry neologism. In SI units, the lowercase letter **m** is the symbol for the meter. The symbol for atomic mass is the lowercase letter **u**. Therefore, the SI correct abbreviation for a mass spectral ion or a peak in a mass spectrum should be *u/z* (*u* = unified atomic mass unit; *z* = number of charges on the particle). This term has never been used. The single term ***m/z*** is a symbol, not an abbreviation. The symbol ***m/z*** should never be used with the word **ratio** (i.e., *m/z* ratio) because "ratio" is part of the definition.

It is important to remember that *m/z* XX is a property (an adjective) of an ion or a peak in a mass spectrum. *m/z* XX is not an ion (i.e., [**correct**] "...the peak at *m/z* 91 represents a significant ion..."; [**incorrect**] "...*m/z* 91 is a significant ion...").

In the case of a molecular ion with a single charge ($M^{+\bullet}$ or $M^{-\bullet}$), the ion results from the loss or gain of an electron from or to the neutral molecule (M). In the case of a fragment ion, the ion could be the result of the formation of a positive or negative ion by breaking a chemical bond. Double-charge ions result in an observed intensity at an *m/z* value of half the mass of the ion, whereas most *m/z* values observed in the mass spectrum are equal to the mass of the ion because $z = 1$. ES ions of large molecules are often multiple-charge ions. The difference between two peaks in a mass spectrum should be reported as *m/z* units (i.e., the difference between the peaks observed at *m/z* 300 and 271 is 29 *m/z* units, not 29 mass units or 29 u). The symbol **u** refers to the unified atomic mass unit, and the symbol ***m/z*** refers to the mass-to-charge ratio.

Incorrect: m/z, *m / z*, or **M/z** – when used as a symbol for the mass-to-charge ratio. The symbol ***m/z*** is always italicized, and its elements are NOT separated by spaces. The *m* is always lower case, even if *m/z* starts a sentence. *m/z* is a symbol, not a mathematical formula. The separated presentation and the use of **M/z** has appeared in print, and the symbol ***m / z*** is presented in the latest edition of *The ACS Style Guide* (Dodd, 1997). However, this use is not the accepted style in any of the major mass spectrometry journals.

Questionable: **thomson** – when used as a name for an *m/z* unit or increment. This neologism was proposed by R. Graham Cooks and Alan L. Rockwood (Cooks, Rockwood, 1991) in honor of Sir Joseph John Thomson as a term to aid in alleviating the confusion caused by the increased occurrence of multiple-charge ions in mass spectrometry. Unfortunately, the term **Thomson number** has already been assigned by IUPAC for use in fluid dynamics.

There are also the physics terms: **Thomson scattering** (named after J. J. Thomson) and **Thomson effect** (a.k.a. Kelvin effect) with a **Thomson coefficient** (μ) named after Sir William Thomson (Lord Kelvin, 1824–1907). However, it appears that the use of **thomson** as a synonym for an *m/z* unit has caught on with a fringe faction in the mass spectrometry community. It is listed as "may be used" by *Rapid*

Communications in Mass Spectrometry. The Cooks/Rockwood letter (as well as the reference in *Rapid Commun. Mass Spectrom.*) does err in the use of Thomson rather than **thomson** and Dalton rather than **dalton** (U.S. Government Printing Office, 1986). The proposed **Th** abbreviation is consistent with rules for abbreviations.

There has been a reluctance by many to accept the use of thomson because it is applying a "unit" to a dimensionless number, which is what *m/z* is by definition.

Incorrect: **m/e** – when used as a symbol for the mass-to-charge ratio when referring to the mass in mass units (u) and the number of charges on the ion. The use of **m/e** would be correct if the mass is reported in kilograms and the charge is in coulombs, which is the case in some mass spectrometry presentations that deal with ion physics. The **m/e** term was the nomenclature used in mass spectrometry prior to 1980 for mass in u and charge number. This usage of m/e is now considered to be archaic. The primary reason for the change is that "e" is the symbol for the charge on a single electron (1.6×10^{-19} columns). This value is not what you want to divide into the mass of the ion! Spectra are seen in older literature with the *x* axis labeled m/e. In this older literature, m/e should be considered synonymous with *m/z*.

Correct: **molecular ion ($M^{+\bullet}$ or $M^{-\bullet}$)** – describes ions in a mass spectrometer or peaks on a spectrum that result from the ionization of a molecule through the gain or loss of electrons. A molecular ion has the same integer mass as the molecule from which it was formed. A spectrum with a molecular ion peak has an observed intensity at the *m/z* value that corresponds to the mass of the analyte molecule divided by the number of charges on the ion. To those who will say, "What about the molecular ion of nitric oxide?": Nitric oxide is not a molecule—it is a radical. There are ions, radicals, and molecules, which are all differentiated from one another. Even though an ion produced by the loss or addition of an electron from a radical has the same mass as a radical, it is not a molecular ion because the resultant ion is not from a molecule. There can be multiple molecular ions of different masses produced for an analyte because of the multiple possible combinations of isotopes.

Incorrect: **molecular ion** – when used to describe the adduct ion produced from a molecule and an ion (e.g., $[M + H]^+$, MH^+, $[M + Na]^+$, etc.).

Correct: **oxonium ion** – an even-electron ion that is the product of a single-bond cleavage of an odd-electron ion that contains oxygen, where the original site of the charge is on the oxygen atom. The oxonium ion has the form: R_2–C=O^+–R'.

*Incorrect: **parent ion ($P^{+\bullet}$)** – when used to refer to the molecular ion. The use of this term to refer to the molecular ion was discontinued when the technique of MS/MS was developed. After that time, parent ion was used to describe the ion (either a molecular ion or a fragment ion) formed in a primary ionization (EI, CI, FAB, etc.) that was selected for secondary fragmentation (dissociation) to produce secondary fragments called daughter ions. These daughter ions could undergo further-induced dissociation to produce granddaughter ions. Because of the gender-specific and anthropomorphic nature of those terms (and the fact that the sex of an ion cannot be determined even when they are turned over), the terms **parent**, **daughter**, **granddaughter**, etc. have been replaced with **precursor** and **product**, respectively.

*Incorrect: **parent**, **daughter**, **granddaughter**, etc. – when used to refer to ions in MS/MS experiments. It has been stated by Gary Glish (Glish, 1992) that the terms **parent** and **daughter** used to describe ions in MS/MS probably had its origin in the analogy with the parent/daughter terminology used to describe nuclear disintegration and other similar relations. He further states, "This terminology is quite appropriate based on definitions in Webster's dictionary, achieves the goal of conveying the desired information in succinct and concise manner, and the genealogy relationship should be easy for scientists (and nonscientists) outside the field [of mass spectrometry] to grasp." However, he agrees with an editorial by Maurice M. Bursey (Bursey, 1992), which states that the term **daughter ion** is found to be offensive to some mass spectrometrists, and "whoever continues to use a term after learning that it is offensive is rude." Jeanette Adams, Emory University, Atlanta, Georgia (Adams, 1992), states that daughter, granddaughter, and great-granddaughter are "archaic gender-specific terms," and that parent ion and progeny fragment ions are "anthropomorphic." She says that because ions are "things" and are incapable of either sexual or asexual reproduction, they can neither be mothers, fathers, daughters, or sons. She further quotes *The ACS Style Guide* (Dodd, 1986): "...discourages the use of gender-specific language in ACS publications (pp 103–105)." It is obvious that she and many others find all these terms to be offensive; therefore, they should not be used.

The *Current IUPAC Recommendations* and the *ASMS Guidelines* state that **product ion** is synonymous with daughter ion, and **precursor ion** is synonymous with parent ion. The gender-specific terms **daughter ion** and **parent ion** are not considered correct by this author. In China, the ions referred to as daughter ions are called son ions because of the importance of the male progeny in the Chinese culture.

*Correct: **precursor ion** and **product ion** – terms that discuss the ions in an MS/MS experiment. The *Journal of the American Society for Mass Spectrometry* lists these two words in "Standard Definition of Terms Relating to Mass Spectrometry" (Price, 1991) as being synonymous with parent ion and daughter ion, respectively. However, in the same issue, Gary Glish (Glish, 1992) suggests the use of parent ion and product ion to describe these two species. Because the terms **parent ion** and **daughter ion** are considered to be offensive by some people, the terms **precursor ion** and **product ion** should be used. Referring to the products of a **product ion** (granddaughter), the term **X-generation product ion** should be used. The *Current IUPAC Recommendations* and the *ASMS Guidelines* state, "This term [precursor ion] is synonymous with parent ion." This synonymous relationship is not considered correct by this author.

Correct: **protonated molecule** – an adduct formed by combining a proton (H^+) and a molecule (M) to give MH^+ or $[M + H]^+$. A protonated molecule is the product of an ion/molecule reaction that has resulted in a positive ion that has a charge of 1 and a mass 1 u greater than the molecule. A protonated molecule can have more than a single charge that results from the addition of more than a single proton. A protonated molecule is an even-electron ion.

When representing a protonated molecule, use MH^+, $(M + H)^+$, or $[M + H]^+$. Do not use $[M+H]^+$ (no space between the three bracketed elements). The same convention is true for other ions that are formed by adding inorganic ions to a molecule (i.e., $[M + Na]^+$, $[M + K]^+$, etc.) and for ions produced by a hydride abstraction ($[M – H]^+$) or that are formed by the loss of hydrogen ($[M – H]^-$). Omission of the spaces between the mathematical sign on the letters can lead to confusion. Does M–H mean a molecule less a proton or a molecule with an added proton?

Some ideas have been suggested to deal with **cationization** of molecules (Bursey, 1992). Generally, these rules involve the use of the stem word used in English for the added element (i.e., sodiuation as opposed to natriation, or silveration as opposed to argentation). However, even the term **cationization** is somewhat questionable with the recommendation that **cation** not be used to describe positive ions in mass spectrometry.

Incorrect: **protonated molecular ion** – when used to describe a molecule that has been protonated. This term implies that the molecular ion (a positive-charge species) has reacted with a proton (another positive-charge species). For this reaction to happen, the two species would have to come in contact with one another. This event is unlikely because like charges repel one another.

Correct: **positive ions** and **negative ions** – terms that describe the charge state of an ion. A **positive ion** is an atom, radical, molecule, or part of a molecule that has one or more fewer electrons than it has protons. A **negative ion** is an atom, radical, molecule, or part of a molecule that has one or more electrons than it has protons. According to the *Current IUPAC Recommendations*: In mass spectrometry, **negative ions** should not be referred to as anions; and **positive ions** should not be referred to as cations because of the connotations of these two terms in solution chemistry. The use of the term **mass ion** is also not considered correct in the *Current IUPAC Recommendations*.

Incorrect: **positive-** or **negative-charge** – when used to refer to the charge on an ion. DO NOT use positive- or negative-charge to describe the ion. Refer to ions as **positive ions** or **negative ions**—the sign of the charge. Ions are charged particles.

Correct: **principal ion** – a term usually reserved to describe ions that have been artificially isotopically enriched in one or more positions. These ions may be either molecular or fragment ions. Examples of principal ions would be $CH_3{}^{13}CH_3{}^{+\bullet}$ (a molecular ion having a mass of 31) and CH_2D^+ (a fragment ion having a mass of 17). The exact definition of a principal ion in the *Current IUPAC Recommendations* is "…a molecular or fragment ion which is made up of the most abundant isotopes of each of its atomic constituents." For the case of species that are not the result of an artificial isotopic enrichment, the principal ion could be another description of a **monoisotopic** or **nominal mass ion**; or in the case of high-mass ions, the **most abundant ion** in an isotope cluster.

Correct: **radical ion** or **odd-electron ion** ($OE^{+\bullet}$) – two terms that can be used interchangeably to describe ions that contain an unpaired electron; therefore, these ions are a radical and an ion. In mass spectrometry literature, the *Current IUPAC Recommendations* states that these ions are represented by placing a superscript dot following the superscript symbol for the charge (i.e., $C_2H_5^{+\bullet}$ and $SF_6^{-\bullet}$). These ions formerly were represented as $X^{+\bullet}$. Placing the dot below the + sign, however, could lead to confusion when there are multiple charges (i.e., $X^{(2+)\,(2\bullet)}$). The *IUPAC Compendium of Chemical Terminology* (1987) lists an alternate form where the dot precedes the sign. This type of presentation is used in the organic and inorganic literature.

Incorrect: **radical cation** or **radical anion** – when used as a synonym for a **positive ion** or a **negative ion**, respectively. Although these terms have been used for years in mass spectrometry, they are no longer recommended by IUPAC (see **positive ions** and **negative ions**, above).

Correct: **rings plus double bonds** – an expression used to describe a method of determining the number of rings and/or double bonds in an ion. The rings-plus-double-bonds calculation results are also known as the **degrees of unsaturation** and **hydrogen-deficiency equivalents**. However, in mass spectrometry, the rings-plus-double-bonds terminology is the preferred:

$$R + db = X - 1/2Y + 1/2Z + 1$$

where X is the number of C and Si atoms, Y is the number of H and halogen atoms, and Z is the number of N and P atoms. Atoms of O and S do not enter into the determination. Double bonds associated with the higher valence states of P and N (valence = 5) or S with a valence of 4 or 6 are not determined with this equation.

*Correct: **single-**, **double-**, **triple-**, or **multiple-charge** – terms used as adjectives to describe the number of charges on an ion. They are expressed as hyphenated words. For example, "The mass spectra of aromatic hydrocarbons can exhibit peaks that represent double-charge ions." This description means the ions have two charges.

*Incorrect: **multiple-charged**, **multiply-charged**, or **multiply charged** or words used to specify the number of charges like **single-charged**, **doubly-charged**, **or triply charged** – when used to describe ions with multiple charges. Ions are charged particles; they are not charged (i.e., a charge is not added to the ion). A charged ion is the same as having a negative deficit. The number and sign of the charge can be changed, but to be an ion means to have an electrical charge. To say an ion is "doubly charged" means an ion was charged once (not likely, because only the sign and the magnitude of the charge on an ion can be changed); then it was charged a second time. These events would result in a particle with three or more or less charges, depending on the sign of the ion's original charge and the sign of the subsequently added charges. The *Current*

IUPAC Recommendations uses singly-, doubly-, triply-charged, etc. This usage is considered to be grammatically incorrect by this author.

*Incorrect: **quasi-molecular ion** – when used to describe an ion that represents the intact molecule. Maurice M. Bursey said, "Show me a quasimolecule, and then I shall agree you can ionize it" (Bursey, 1991). This term was originally used to indicate an ion with the approximate, but not the exact, mass of the molecule. An example would be the $[M + 1]^+$ formed in the protonation of a molecule under chemical ionization conditions. The term was later used to describe any ion formed by a process that did not generally disturb the structure of the molecule. In **positive-ion mass spectrometry**, examples would be $[M + 1]^+$, $[M - 1]^+$, $[M + Na]^+$, etc.; and in **negative-ion mass spectrometry**, $[M - 1]^-$, $[M + Cl]^-$, etc. The use of quasi-molecular ion is, however, recommended by IUPAC.

Incorrect: **pseudo-molecular ion** – when used to describe the same type of ion as described by quasi-molecular ion, and has the same degree of ambiguity. Both of these terms fall into the class of words "that mean what the user wants them to mean." Therefore, they can mean different things to different users.

*Correct: **resolving power** – the term used to define the ability of a mass spectrometer to separate ions of two different *m/z* values. The resolving power of the mass spectrometer is defined as **M/ΔM** where **M** is the *m/z* value of a single-charge ion and **ΔM** is the difference between M and the value of the next highest *m/z* value ion that can be distinguished (separated) from M in *m/z* units. Resolving power is determined from the measurement of mass spectral peaks and should be reported with the method by which ΔM was determined (i.e., full width at half-maximum height [FWHM], 10% valley [two adjacent mass spectral peaks of equal height overlap so that the height from the baseline to the top of the overlap valley is 10%] method, 50% valley method, etc.; the FWHM results in a value for resolving power that is twice that obtained with the 10% valley method. It should be noted that using the mass spectral peak width at 5% of its full height is considered to be equivalent to the 10% valley method as illustrated in **Figure 4**).

*Incorrect: **resolution** – when defined in the same way as **resolving power**. Resolution is the inverse of resolving power and expressed as **ΔM** at **M**. Although resolving power is a large number and is associated with a "valley" or

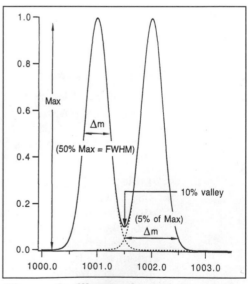

Figure 4. Illustration of the peaks used to calculate the resolving power of a mass spectrometer showing the location of the FWHM, the 10% valley, and the 5% valley.

mass spectral peak width, resolution can be a small number. Resolution is the measure of a separation of two mass spectral peaks. Resolution is often reported in terms of parts per million (ppm). In physics, resolution is used to describe the separation of a vector into its components or the amount of information or detail revealed in an image procedure. In chemistry, resolution is used to describe the separation of a "racemic mixture" into its optically active components or components in a chromatographic separation.

Note: The *Current IUPAC Recommendations* and *ASMS Guidelines* define **resolution** and **resolving power** in the opposite way than how they are defined in this document. However, the "MS Terms and Definitions" that appears on the ASMS Web site (http://www.asms.org) under Items of Interest uses language that is in agreement with the above definitions.

Incorrect: **satellite ions** or **satellite peaks** – when used as synonyms for **isotope** or **isotopic**, respectively. Because a satellite usually "revolves" or "is" around or in some way associated with some primary object, the use of satellite to describe peaks in a mass spectrum or ions related to an ion of a given *m/z* value could refer to [X – 1], [X – 2], etc. as well as those that are possibly involved with isotopes [X + 1], [X + 2], etc. This ambiguity is the reason why the word **satellite** should not be used when referring to isotopes.

Correct: **tropylium ion** – an even-electron ion that is the product of a single-bond cleavage of an odd-electron ion that contains a benzyl group (C_6H_5–CH_2–R) and the charge is on the ring. The tropylium ion results from a cleavage between the methylene carbon and the atom adjacent to it followed by the formation of a seven-member ring with a phenyl group and the methylene carbon.

A Tropylium Ion

Correct: **tolyl ion** – an even-electron ion that is the product of a single-bond cleavage of an odd-electron ion that contains a tolyl group (CH_3–C_6H_4–R). The tolyl ion results from a cleavage between the R group and the ring to produce an ion with the form: CH_3–$C_6H_4^+$ where the charge is on the ring.

CH_3

A Tolyl Ion

Note: The word **isotope** modifies the word **peak**, **cluster**, and **number**. The word **isotopic**, which is an adjective, modifies the word **ion** (author's convention).

MASS

Correct: **atom** – the fundamental particle of a chemical element. Atoms are neutral. They are composed of a nucleus that consists of positively charged particles (protons) and neutral particles (neutrons), both having about the same mass. The **nominal mass isotope** of hydrogen is the only atom that does not contain neutrons. There are a number of electrons, which are equal to the number of protons, that exist in energy shells (orbitals) around the nucleus. Electrons have a mass that is 1/1837 that of a proton. Atoms are characterized by their **atomic number** and **mass number**. Atom is derived from the Greek word *átomos*, which means undivided. The principle of the atom and the word was first proposed and used by the Greek philosopher, Democritus (470–380 BC).

Correct: **u** – the symbol for a mass unit or the *unified atomic mass unit*. This symbol represents 1/12th the mass of the most abundant naturally occurring stable isotope of carbon. The term **dalton** (symbol **Da**) is used in biochemistry and is accepted in mass spectrometry; however, it has not been approved by the Conférence Générale des Poids et Mesures. The mass of a given particle is the sum of the atomic masses (in daltons) of all the atoms of the elements composing it. The use of the symbol **u** is referenced in *Quantities, Units and Symbols in Physical Chemistry*, 2nd ed. (Mills, 1993).

Incorrect: **amu** – when used as the symbol for mass unit. The *Current IUPAC Recommendations* no longer recommends amu as the symbol for mass. The amu symbol for a unit of mass was used when the standard for mass was based on "oxygen 16." Physicists reported mass based on the most abundant naturally occurring stable isotope of oxygen (^{16}O: established by Francis William Aston [1877–1945] in 1929 after his discovery that oxygen was composed of three different isotopes, two of which had a higher mass [^{17}O and ^{18}O] than the most abundant isotope), which was assigned an exact mass of 16 (1 amu = 1/16 of the mass of ^{16}O). This definition was the basis of the *physical atomic mass scale*. Chemists used amu to define a unit of mass as 1/16 the atomic weight (the average atomic mass) of oxygen (officially established in 1905 on the suggestion of the Belgium chemist, Jean Servais Stas [1813–1891]). This definition was the basis for the *chemical atomic mass scale*. The two scales differed by a factor of 1.000275 (physical > chemical).

An **atomic weight** is the weighted average of the masses of the naturally occurring stable isotopes of an element, and oxygen has three such isotopes. The atomic weight of oxygen was an absolute value of 16 by definition. The chemical atomic mass scale made the determination of the atomic weight of newly discovered elements easy by forming their oxides.

To eliminate the ambiguity between the physical and chemical standards, the standard of a single mass unit as 1/12 the most abundant naturally occurring stable isotope of carbon (^{12}C) was adopted in 1960 by the International Union of Physics at Ottawa and in 1961 by the International Union of Chemists at Montreal. This standard is based on the independent recommendations of D. A. Olander and A. O. Nier in

1957. The symbol for the *unified atomic mass unit* was established as **u**, NOT µ, which appears in the *ASMS Guidelines* and in the 4th edition of *Interpretation of Mass Spectra* (McLafferty, 1993).

Prior to the oxygen standards, the basis for atomic mass had been set ca. 1805 by John Dalton (1766–1844) as 1 for the lightest element, hydrogen. In 1815, the Swedish scientist, Jöns Jacob Berzelius (1779–1848), set the atomic weight (relative atomic mass) of oxygen to 100 in his table of atomic masses; however, the Berzelius standard of mass was not accepted. Stas' recommendation of setting the "oxygen 16" standard allowed hydrogen to retain a mass close to 1, thereby keeping Dalton's scale somewhat intact.

Inappropriate: **A.M.U.** or **a.m.u.** – when used as abbreviations for atomic mass unit. However, they should be avoided because of the confusion with the symbol **amu** (see **amu**, above). The symbol for the unified atomic mass unit is **u**.

Note: The use of **amu**, **a.m.u.**, **A.M.U.**, or any other term used for mass (**dalton**, **Da**, **u**, or **mass**) as a label for the abscissa of a mass spectrum is incorrect when the abscissa represents values of mass-to-charge ratios of ions. Only use terms that represent units of mass when the *m/z* values have been converted to mass; never use amu, a.m.u., or A.M.U. to represent "mass."

In the older literature, the abscissa of the mass spectrum was often labeled "mass." This type of presentation was used because, in nearly every case, the number of charges on the ion was one. After the development of the electrospray interface and the analysis of biomacromolecules with the presence of multiple-charge ions became widespread, the use of "mass" on the abscissa of the mass spectrum (although always ill advised) became more problematic and potentially confusing.

Care must also be taken in the use of the terms **molecular weight** and **molecular mass**. The **molecular weight** (a.k.a. relative molecular mass) of a molecule is based on the **atomic weights** (relative atomic masses to several decimal places) of its atoms, whereas the **molecular mass** is based on either the **nominal**, **monoisotopic**, or **isotopic masses** of its elements.

Correct: **atomic number** – the number of protons in an atom of a element. All elements have a unique atomic number that determines the position of the element in the periodic table.

Correct: **average mass** – the calculated mass of an ion based on the atomic weights (relative atomic mass) of the elements from which it is composed. An **atomic weight** is the weighted average of the naturally occurring stable isotopes of an element. Atomic weight of carbon:

$$(12.000 \times 0.989) + (13.0034 \times 0.011) = 12.0110 \text{ (see } \textbf{Figure 5})$$

Incorrect: **atomic weight** – when used to describe the mass of individual isotopes of an element. The integer mass of the most abundant naturally occurring stable isotope of an elemental is the **nominal mass**. The exact mass of the most abundant naturally occurring stable isotopes of an element is the **monoisotopic mass**.

Correct: **calculated exact mass** – the mass determined by summing the mass of the individual isotopes that compose a single ion, radical, or molecule based on a single mass unit being equal to 1/12 the mass of the most abundant naturally occurring stable isotope of carbon. The exact mass values used for the isotopes of each element are recorded in tables of isotopes. The mass (atomic weight) of an element that appears in the periodic table is a weighted average of these exact mass values of the naturally occurring stable isotopes of that element. If the mass is calculated with the exact mass value of the most abundant naturally occurring stable isotope of each element in the ion, radical, or molecule, then the **calculated exact mass** is the same as the **monoisotopic mass**.

Incorrect: **exact mass** – when used to mean the experimentally determined (measured with a mass spectrometer) mass of an ion, radical, or molecule. Exact mass has been used to describe both measured and calculated mass, leading to a degree of confusion. Exact mass should never be used to mean an accurately measured mass.

Correct: **measured accurate mass** – an experimentally determined mass of an ion, radical, or molecule that allows the elemental formula to be deduced. For ions with a mass ≤ 200, a measurement to ± 5 ppm is sufficient for the determination of the elemental composition. The term **measured accurate mass** is used when reporting the mass to some number of decimal places (usually a minimum of 3). A measured mass should be reported with a precision of the measurement (Gross, 1994).

Incorrect: **accurate mass** – when used to mean the calculated mass of an ion, radical, or molecule. Accurate mass has been used to describe both measured and calculated mass, leading to a degree of confusion. Accurate mass should never be used to mean calculated mass. The use of accurate mass in the reporting of a measured mass should be used with the word **measured**—a **measured accurate mass**.

Correct: **isotopes** – atoms of the same element that differ in **mass number** (the sum of the protons and neutrons that compose the nucleus of the atom) due to the difference in the number of neutrons in the nucleus. The term was proposed by Frederick Soddy (1877–1956), the English physical chemist, in 1917, for different radioactive forms of the same chemical species because they are classified in the same place in the periodic table (Gr, *isos*, equal, same +, Gr, *topos*, place; that is, "having the same place" in the periodic table). Later, Francis William Aston (1877–1945), an English physicist, applied the term to atoms of the same element that had different mass but that did not undergo radioactive decay—stable isotopes. Aston is best known for the discovery of most of the naturally occurring stable isotopes of the first 92 elements in the periodic table. Both Aston and Soddy were Nobel Prize recipients and worked in J. J. Thomson's Cavendish Laboratory.

Correct: **isotope number** – the difference between the number of neutrons and the number of protons in a nuclide. Different isotopes of the same element have different **isotope numbers**.

Correct: **isotopic mass** – the mass of any isotope.

Correct: **mass defect** – in mass spectrometry, the difference between the exact mass of an atom, molecule, ion, or radical and its integer mass.   In physics, the mass defect is the amount by which the mass of an atomic nucleus is less than the sum of the masses of its constituent particles.

Correct: **mass number** – the sum of the total number of protons and neutrons in an atom, molecule, ion, or radical.   It is the **nucleon number** with the symbol **m**.   This number is an integer and can be used interchangeably with *m/z* values in unit-resolution mass spectra where the charge number of the ion is one.   However, this practice is not recommended because of the possible confusion that may result.   The mass number of isotopes of two different elements can be the same.   Although the two elements would each have a unique number of protons, the number of neutrons in each element could be such that both elements had the same mass number.   Zirconium has an **atomic number** of 40, and molybdenum has an **atomic number** of 42; however, both have an isotope with a **mass number** of 92 (^{92}Zr and ^{92}Mo).

Correct: **molecules** – groups of atoms that are chemically bonded to one another.   Molecules have no charge and are characterized by an even number of electrons.   The bonds can be covalent (the sharing of a pair of electrons) or ionic (attraction of opposite charges).   Particles that have an odd number of electrons that do not have a charge are radicals (nitric oxide (NO) is a radical, not a molecule).

Correct: **most abundant mass** – represented by the most intense peak in an isotope cluster for an ion (e.g., the [X + 2] peak for an isotope cluster that represents an ion that contains only carbon, hydrogen, and four atoms of chlorine).

Correct: **monoisotopic mass** – the exact mass of the most abundant naturally occurring stable isotope of an element.   The monoisotopic mass of an ion is the sum of the monoisotopic masses of the elements in its empirical formula (e.g., $C_3H_6O^{+\bullet}$ has a monoisotopic mass of 58.0417).   The monoisotopic mass of an element is not necessarily the lowest mass naturally occurring stable isotope.   For the common elements in organic mass spectrometry (C, H, O, N, S, Si, P, and the halogens), the lowest mass isotope is the monoisotopic mass isotope (see **Figure 5**).

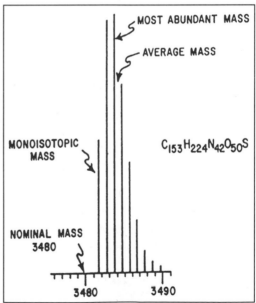

Figure 5.   Illustration of the relationship among the nominal mass, monoisotopic mass, most abundant mass, and average mass of a large-mass single-charge ion. (Yergey, 1983)

Correct: **molecular mass** – describes the mass of a molecule or molecular ion. This term should not be used to describe the mass of a fragment ion, adduct ion, or radical. The relative molecular mass (M_r) is based on atomic weights of the elements of the molecule.

Correct: **neutral loss** – describes a radical or molecule that is lost from an ion to produce an ion of lower mass. Neutral losses should be reported as units of mass because they represent species that have no charge.

Correct: **nitrogen rule** – Any common organic molecule or odd-electron ion ($OE^{+\bullet}$) [neither an even-electron ion (EE^+) nor radical], containing C, H, O, S, P, and/or halogens, that has an odd number of nitrogen atoms has an odd nominal mass.

Correct: **nominal mass** – the integer mass of the most abundant naturally occurring stable isotope of an element. The nominal mass of an ion is the sum of the nominal masses of the elements in its empirical formula (e.g., the nominal mass of $C_3H_6O^{+\bullet}$ is 58). The nominal mass of an element is usually equal to the **mass number** of the most abundant isotope of an element. The lowest mass isotope is not necessarily the nominal mass isotope.

The lowest mass isotope of Hg is 196 (0.1%). The nominal mass is 202 (29.52%). Hg has seven naturally occurring stable isotopes: 196, 198, 199 (16.8%), 200 (23.1%), 201, 202, and 204.

The lowest mass isotopes for the common elements in organic mass spectrometry (C, H, O, N, S, Si, P, and the halogens) are the nominal mass isotopes (see **Figure 5**).

Incorrect: **nominal mass** – when defined as the integer mass of any isotope of an element or as the integer mass of the lowest mass isotope of an element. Both of these definitions have appeared in mass spectrometry literature. It is important to remember that the integer value for the monoisotopic mass of an ion is NOT the nominal mass of that ion [e.g., the monoisotopic mass of $C_{153}H_{224}N_{42}O_{50}S$ is 3481.59985; its nominal mass is 3480, NOT 3482; the nominal mass is the sum of 153(12), 224(1), 42(14), 50(14), and 1(32)].

Correct: **nuclide** – an atom characterized by its atomic number, mass number, and nuclear energy state. Nuclide can be considered as a synonym for isotope.

Correct: **radical ($X^{\bullet}$)** – a neutral particle that has an unpaired electron (an odd number of electrons).

Incorrect: **relative** or **average atomic weight** – when used to describe the weighted average mass of the naturally occurring stable isotopes of an element.

Correct: **whole number rule** – the mass of all atoms (the different naturally occurring stable isotopes of all elements) can be expressed to a high degree of accuracy (in most cases to about 1 part in a thousand). The error expressed in fractions of a mass unit increases with increasing mass. The one exception is hydrogen with an exact mass of 1.007825. There are no atoms of Cl with a mass of 35.6 (the atomic weight of Cl), only those that have a mass of 34.9689 (^{35}Cl) and 36.9659 (^{37}Cl). The difference in the whole number and the actual mass is the **mass defect**. The **whole number rule** was put forth by the British chemist William Prout in 1815 and vindicated by F. W. Aston in 1920 (Aston, 1924).

SOME OTHER IMPORTANT DEFINITIONS

Note: There are several important words that can create a great deal of confusion in GC/MS, LC/MS, and mass spectrometry because of their double or multiple meanings. The following are the multiple definitions for some of these words:

calibration –   mass calibration done by the data system to assign digital-to-analog conversion (DAC) values to mass spectral peaks of some substance that produces ions of known *m/z* values; calibration of the *m/z* scale of the mass spectrum.

  –   a plot of target analyte amounts versus chromatographic peak area or height, response factor, etc.; NOT mass calibration.

tuning –   adjustment of the mass spectral peak shape, and the resolution between peaks of two adjacent *m/z* values.

  –   adjustment of the relative intensities of mass spectral peaks during the analysis of compounds such as decafluorotriphenylphosphine (DFTPP) (e.g., those at *m/z* 198 and 442).

peak –   **mass spectral peak** – produced by the energy distribution of ions of a given mass-to-charge ratio that strikes the detector of the mass spectrometer.

  –   **chromatographic peak** – produced by plotting the intensity of signals corresponding to ions of different *m/z* values in consecutive spectra that were acquired during the elution of a compound. These intensities may be directly related to the concentration of the analyte during ionization, depending on whether or not there is coelution.

TIC –   In GC/MS and LC/MS, **TIC** (total-ion chromatogram) is often used incorrectly to describe a **reconstructed total-ion-current chromatogram.**

  –   TIC is also used in U.S. EPA methods to describe **tentatively identified compounds**.

  –   TIC can also be used to describe the **total ion current** in a mass spectrum or display.

library –   a collection of mass spectra used to compare against a spectrum of an unknown compound. Mass spectral libraries (a.k.a. databases) are used in mass spectral library searches.

  –   a database of proteins that is a listing of the amino acids of each protein in the library. These libraries DO NOT contain mass spectra.

deconvolution –   the separation of coeluting components, or of spectra, in a reconstructed chromatographic peak.

  –   the determination of the mass of an ion based on the mass spectral peaks that represent multiple-charge ions (charge deconvolution).

Correct: **sensitivity** – is actually the slope of a plot of analyte amount versus signal strength. Unfortunately, the term is incorrectly used by many instrument manufacturers to pertain to the amount of analyte put into the primary sample inlet of an instrument that produces an "acceptable" signal (more accurately: the **detection limit**). The "sensitivity specification" of an instrument should contain the instrument conditions and a defined result (e.g., in GC/MS, 2 pg of hexachlorobenzene of a sample injected onto a GC column will produce a mass chromatographic peak of *m/z* 282 with a signal-to-background of 10:1 and a spectrum identifiable as hexachlorobenzene when searched against the NIST/EPA/NIH Mass Spectral Main Library). Important instrument conditions for a GC/MS "sensitivity specification" are: GC column (length, diameter, stationary phase and its thickness), linear velocity of carrier gas, method of column interface with mass spectrometer (jet separator, open-split interface, direct to ion source, etc.), GC-column oven-temperature program rate, mass spectral acquisition rate, ionization energy, and detector gain (a new detector can be operated at a higher gain to get better sensitivity, but will not represent typical operation because detector life is diminished and noise rapidly increases under these conditions).

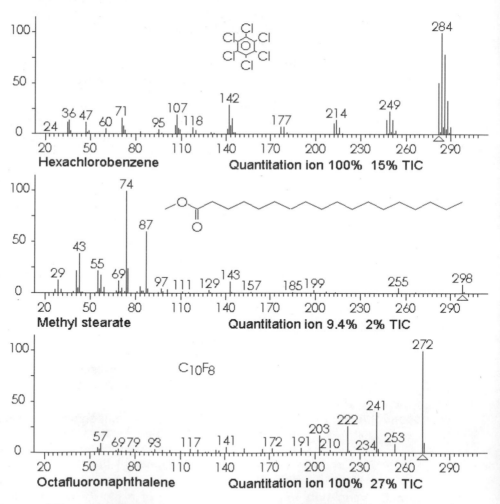

Figure 6. Mass spectra of three compounds that were used to demonstrate the sensitivity in GC/MS instrumentation.

Care must be taken with respect to the evaluation of a "sensitivity specification." **Figure 6** shows the mass spectra of three different compounds used to demonstrate the sensitivity of a GC-MS. The percent of ion current represented by the ion used to produce the mass chromatographic peak differs in each case. It is easy to see that a "sensitivity specification" of 1 pg of octafluoronaphthalene is equivalent to a "sensitivity specification" of 2 pg of hexachlorobenzene.

Another factor that has to be taken into account in evaluating a "sensitivity specification" is the presence of chemical background. The use of hexachlorobenzene as a detection-limit standard can be complicated by the [X + 1] ion associated with the column-bleed ion at m/z 281. The m/z 282 isotope from this GC column bleed will significantly contribute to background and will result in a reduction of the signal-to-background value.

The *Current IUPAC Recommendations* defines sensitivity as: "Two different measures of sensitivity are recommended. The first method, which is suitable for relatively involatile materials as well as gases, depends upon the observed change in ion current for a particular amount or change in flow rate of the sample through the ion source. The recommended unit is a coulomb per microgram ($C \; \mu g^{-1}$). A second method stating sensitivity that is most suitable for gases depends upon the change of ion current related to the change of partial pressure of the sample in the ion source. The recommended unit is ampere per pascal ($A \; Pa^{-1}$)." However, sensitivity is not defined this way by the instrument manufacturers. The *Current IUPAC Recommendations* does differentiate between sensitivity and detection limit, and does specify the importance of the specification containing the experimental conditions.

Correct: **signal-to-background ratio (S/B)** – the measure of the signal strength for a sample compared to a region with no sample during a single measurement. S/B is often incorrectly called signal-to-noise. An example of signal-to-background in GC/MS is the measurement of the height of a mass chromatographic peak compared to the width of the baseline, disregarding the zero offset.

Correct: **signal-to-noise ratio (S/N)** – the measure of the noise in the presence of the signal. S/N is the variation of several subsequent measurements of peak height (mass spectral or chromatographic) for the same amount of sample introduced. The variation is the noise, and the peak height is the signal. **Signal-to-background** is a measure of detectability. **Signal-to-noise** is a measure of precision.

Correct: **precision** – the variation within a single measurement (noise) or from measurement to measurement, or the degree of significant figures used to report a number. Precision is related to, but different from, accuracy. Precision indicates degree of detail; accuracy indicates correctness. The number 3.1416 is more precise than 3.14, but is less accurate for pi.

Incorrect: **precision** – when used to describe the difference between calculated and measured mass.

Correct: **accuracy** – the difference between the actual mass (based on literature values for the individual isotopes) and the measured mass.

Correct: **background** – the term used to describe a signal observed in a mass spectrum that comes from the presence of chemical materials other than the analyte. This background could be due to GC column bleed (usually cyclic alkyl-siloxanes produced by the thermal decomposition of the column's stationary phase, which have molecular weights <500 that are characterized by mass spectral peaks of decreasing intensity at m/z 207, 281, 355, and 439). Background signal can also come from noise in the electronics.

Correct: **background subtraction** – the process of subtracting the absolute intensity of each mass spectral peak in a background spectrum from the absolute intensities of each mass spectral peak in a sample spectrum. The background spectrum and the sample spectrum can be a single spectrum or an average of a series of spectra. Although this subtraction can result in negative intensity values for peaks, the resulting **background-subtracted spectrum** displayed will not have negative peaks. Background subtraction can result in a spectrum that is easier to interpret because it is free from peaks that represent contaminants.

When a reconstructed total-ion-current chromatographic peak represents more than one analyte, a spectrum representing the preponderance of a single analyte can be subtracted from the spectrum that represents the preponderance of a second analyte (not necessarily the spectrum representing the apex of the chromatographic peak) to obtain a pure spectrum of the second analyte. The reverse subtraction can then be carried out to obtain a pure spectrum of the second analyte. Care must be taken in that a significant peak in one of the mass spectra is not eliminated by the process when the analytes have ions of common m/z values.

Correct: **space charge** – the term used to describe the cause of the broadening of the mass spectral peak in an ion trap when too many ions are present. Because most mass spectral data are presented in a bar-graph form, the effects of space charge are a series of peaks at higher and lower m/z values, along with peaks for the specific ion. This "Christmas tree"-shaped peak cluster is a result of space charge. Because space charge occurs when the sample pressure is too high, it has been incorrectly referred to as self-CI.

Correct: **ion/molecule reaction** – the term used to describe ALL reactions that take place between ions and molecules in the ion source of a mass spectrometer. In an ion-trap mass spectrometer (quadrupole or magnetic), molecular ions of certain compounds undergo a γ-hydrogen shift-induced rearrangement followed by a beta cleavage to produce a distonic odd-electron ion that is a very good proton donor.

These distonic ions react with analyte molecules of the same analyte to produce protonated species because these types of molecules have high-proton affinities. These ion/molecule reactions have been referred to as **self-CI**. The term **ion/molecule reaction** describes the process that occurs in chemical ionization mass spectrometry.

Self-CI can also mean the chemical ionization of an analyte where the reagent ion is from an ionized form of the analyte itself (i.e., a fragment ion or a molecular ion of the analyte that protonates other molecules of the same analyte, OR analyte molecular ions or EI fragment ions that ionize other molecules of the analyte by charge transfer or addition).

Incorrect: **self-CI** – when used to describe space charge and ion/molecule reactions in a quadrupole ion-trap mass spectrometer. Self-CI is a term first used by Wilkins and Gross (Ghaderi, 1981) to describe the protonation of analyte molecules by fragment ions that result from the electron ionization of the same analyte in an FTICR-MS. Self-CI has been incorrectly used to describe not only this phenomenon that can take place inside an internal ionization quadrupole ion-trap (I^2QIT) mass spectrometer or FTICR-MS but also the space-charge effect, which is mass spectral peak broadening (loss of resolution) brought about by having more ions than can be stored in the trap.

Correct: **SI units** (Système International d'Unités derived from the *m.k.s. system* (meter, kilogram, second), thereby replacing the *c.g.s. system* (centimeter, gram, second) and the *f.p.s. system* (foot, pound, second) for all scientific purposes) – for most units in mass spectrometry.

For example, the use of **pascal** (abbreviated **Pa**: the derived SI unit of pressure, equal to 1 newton [the force required to accelerate a standard kilogram at a rate of 1 m/s^2 over a frictionless surface] per square meter) can be substituted with **torr**[*] (760 Torr is equal to 1 atmosphere [1 atmosphere is the force exerted by the pressure of the atmosphere at 0 °C sea level and 45° latitude on an open reservoir of mercury (density 13.5950 g/cm^3 at 0 °C) so that a column of mercury in a tube with one end sealed and the other end open and immersed in a reservoir of mercury can be supported to a height of 760 mm. 1 atm = (13.5950 g/cm^3)(980,665 cm/s^2)(76.00 cm)]. This procedure is the origin of the expression **centimeters-of-mercury** or **inches-of-mercury pressure**). Because pressure is a ratio of force to area and not length, the **pascal** is more descriptive; however, currently, the term **torr** is more common in mass spectrometry. By definition: 1.013×10^5 Pa = 1 atm = 760 Torr. 1 Pa is $\cong$ 0.0075 Torr, and 1 Torr = 133.322 Pa.

A pressure unit that is often encountered on ion-gauge readouts is the **millibar** (**mbar**), which is the standard unit of pressure in the *c.g.s. system*. A bar is a pressure of 10^6 dynes per sq. cm or 10^5 Pa. The mbar = 100 Pa.

Another non-SI unit found in the mass spectrometry literature is **kcal mol^{-1}** or **eV** (the energy of an electron accelerated by falling through a potential of 1 volt) as the unit of energy instead of the **joule** (the SI unit for energy). The eV molecule^{-1} often appears in mass spectrometry publications.

Note: When a unit is named after a person, the first letter of the abbreviation is upper case. If the name is spelled out, the first letter is lower case (i.e., **Da** and **dalton**). The abbreviation should only be used after a number (U.S. Government Printing Office, 1986).

---

* Although the torr unit of measure is an abbreviation of the name of Galileo's successor, Evangelista Torricelli (1608–1647), it is treated as the entire name and is written with the first letter as lower case when it is used without a number. As a lowercase abbreviation, torr is the one exception to the rule for the use of proper names and their abbreviations as units of measure when not associated with a number.

Note: The SI system of reporting one unit per another unit (e.g., grams per milliliter) is usually written as "first unit second unit^{-1}" (i.e., **µg mL^{-1}** rather than µg/mL). Most publications accept either form.

Note: The use of "gms" or "gm" as the symbol or abbreviation for grams or gram, respectively, is considered archaic. The correct symbol for **grams** or **gram** is the lowercase letter **g**.

Although the lowercase letter **l** was used as the symbol for liter for many years, the accepted symbol is an uppercase **L** (e.g., **mL** instead of ml, or **nL** instead of nl). When referring to microliters, do not use the symbol λ (the Greek lowercase lambda)—use **µL**, not uL!

A good reference for abbreviations, symbols, and general style information is *The ACS Style Guide: A Manual for Authors and Editors*, J. S. Dodd, Ed., first published in 1986; now in its 2nd edition (1997). Even though different journals and editors have some deviations from *The ACS Style Guide*, it is a good resource. Just as a point-in-fact, in *The ACS Style Guide*, 2nd ed., p 170, Table 6: Non-SI Units That Are Discouraged, the term **torr** appears. This difference between the mass spectrometry standards and the standards of the ACS shows the somewhat rebellious nature of mass spectrometrists in regard to nomenclature and terminology.

Correct: **sample** – a mixture of the analyte(s) and a matrix (the material that contains the analyte). The term **sample** can be used to describe the material taken for analysis (i.e., a plasma sample that contains a drug and its metabolites) or the material put into the instrument for the determination of analytes (i.e., the MALDI matrix that contains the analyte).

Correct: **analyte** – the material (usually an individual compound) in the sample that is being determined by the analysis. However, it should be noted that an analyte can represent a mixture, such as the case of the analysis of soil for gasoline. A capillary electrophoresis purification of a sample results in a pure analyte.

Correct: **eluate** – the material that comes from the chromatographic column that consists of the mobile phase, analytes, any dissolved stationary phase, and any components introduced with the sample.

Correct: **eluant** – the mobile phase in a chromatographic process; the substance used in the process of elution. The word **eluant** has also been spelled **eluent**, which is a modern word that first appeared during the World War II era (1940–1945).

Correct: **intramolecular** – involves only the decomposition of individual ions.

Correct: **intermolecular** – involves the reaction of ions with other ions or neutral molecules or fragments.

Correct: **matrix** – the material that contains the analyte(s) that is(are) being determined.

Correct: **mean-free path** – the minimum length that half of all ions present in an *m/z* analyzer will travel before they encounter another ion or molecule.

SOME OTHER TERMINOLOGY TO AVOID

Inappropriate: **moiety** – a word that crept into the mass spectrometry literature of the 1960s from organic chemistry. The publication of many mechanistic rationalizations prompted a search for synonyms of fragment. Moiety usually means portion (i.e., the benzoyl portion of the molecule is the benzoyl fragment or moiety). However, as can be seen from the definition below, its primary meaning is "a half." Maurice M. Bursey (Bursey, 1991) suggests that this word is not a good English word and should not be used.

> moiety (moi¹î-tê) noun
> plural, moieties
> 1. a half
> 2. a part, portion, or share
> 3. either of two basic units in cultural anthropology that makes up a tribe on the basis of unilateral descent
> [Middle English *moite*; from Old French *meitiet, moitie*; from Late Latin *medietâs*; from Latin *medius*, middle]

Correct: **fragment**, **portion**, or **part** – terms used to refer to a portion of a molecule in mass spectrometry (i.e., a benzoyl fragment or loss of the methyl portion of the molecular ion).

Correct: **formula weight** (**mass**) – a term that is generally not used in mass spectrometry. This mass is based on the integer or 0.5 value of an atomic weight (weighted average).

1. There is a tendency in scientific writing to avoid the use of personal pronouns. There is nothing wrong with the use of personal pronouns in technical presentations. Some authorities encourage the use of personal pronouns because it gets the reader more involved. The word "one" is often used as a poor (and awkward) substitute for the personal pronoun and should be avoided.

   Awkward: If one uses gas chromatography/mass spectrometry in the analysis of underivatized amino acids, then one will get little information.

   Correct: If you use GC/MS in the analysis of underivatized amino acids, then you will get little information.

   Preferred: If GC/MS is used in the analysis of underivatized amino acids, then little information will be obtained (*in the event you do not want to use personal pronouns*).

   Either use personal pronouns or don't use them. Do not substitute the word "one" for a personal pronoun.

2. In lecturing, there will be a tendency to use the term "note that." The inclusion of "note that" can be overused in writing, and its excessive use should be avoided. The same is true for "Thus, ..." to begin a sentence.

INSTRUMENTS

Types of *m/z* Analyzers

double-focusing mass spectrometer: This instrument has a magnetic sector (single-letter symbol **B**) and an electric sector (single-letter symbol **E**: a.k.a. electrostatic sector).   This type of instrument is often referred to as a **high-resolution** mass spectrometer because of its ability to separate ions that have very small differences in *m/z* values (0.0001).   The electric sector is used to produce a dispersion of ions based on their kinetic energy, whereas the magnetic sector produces a dispersion of ions based on their momentum.   In a **forward-geometry** instrument (ions pass through the electric sector before the magnetic sector), ions of a narrow kinetic energy pass through a slit to the magnetic sector, where ions with a small difference in mass are separated based on their momentum.   In a **reverse-geometry** instrument (magnetic sector preceding the electric sector), ions of a common momentum enter the electric sector, where they are dispersed based on their differences in kinetic energy.   Only those of a narrow kinetic energy are allowed to pass through a slit into the ion detector, thereby accomplishing the precise mass measurement. Sometimes these instruments are described based on the location of the two fields relative to the ion source and the detector (see **Terms Associated with Double-Focusing Mass Spectrometers**) or according to the names of the developers of their particular ion-optic geometry (i.e., Mattauch–Herzog or Nier–Johnson).

Fourier transform ion-cyclotron resonance (**FTICR**) mass spectrometer: This instrument is a magnetic ion trap (a.k.a. **Penning ion trap**).   The technique of Fourier transform ion-cyclotron resonance mass spectrometry (FTICRMS) in commercially produced instruments is currently limited to the use of **ion-cyclotron resonance** (**ICR**) mass spectrometers.   Some FT ion detection has been experimented within quadrupole ion traps (Badman, 1998).   In FTICRMS, ions are trapped in a cell (which is inside a strong magnetic field: **B**) composed of three distinct sets of plates—trapping, transmitter, and receiving plates.   The ions in the trap move in circular orbits (cyclotron motion) in a plane perpendicular to the magnetic field.   The ions are held in the cell by electric potentials applied to the trapping plates that are perpendicular to the magnetic field.   Mass analysis is accomplished by the application of a radio frequency electric potential to the transmitter plates to cause trapped ions to be excited into larger circular orbits.   As the excited ions pass near the receiver plates, the frequency of their passage is detected as an induced current in the plates called "image" current.   The frequency of the motion of an ion is inversely proportional to its mass.   This type of ion detection results in a nondestructive detection of ions.   The signal-to-background is enhanced by averaging many cycles before transforming and storing data.   It is possible to use lasers to fragment ions in these traps to perform MS/MS.   These instruments are capable of extremely high resolving power.   In order to obtain the desired magnetic field strength, it is necessary to use a super-conducting magnet.

magnetic-sector mass spectrometer: This single-focusing instrument has only a single magnetic sector. In modern instruments, ions are accelerated from the ion source at a fixed value of 1–10 kV into a tube of a fixed radius of curvature that is subjected to a magnetic field with the direction of magnetic flux at a right angle to the path of the ions. Under these conditions, only ions of a single, narrow *m/z* range will have the same radius of curvature as the tube and reach the end of the tube, where a detector is mounted. It is possible to measure all ions of different *m/z* values by incrementally changing the field strength of the magnet or by allowing the ions to impinge on a plane (as opposed to a point) in an instrument with a fixed-field magnet. In this latter case, ions of differing *m/z* values have different radii of curvature and strike the plate in different locations, according to their different *m/z* values. It is also possible to use a fixed-field magnet and incrementally change the accelerating voltage to obtain a mass spectrum. This technique is often employed in selected ion monitoring with sector-based instruments. These instruments cannot produce much better than unit to 0.1 *m/z* resolution. The equation used to describe ion separation in a magnetic-sector instrument is:

$$m/z = R^2 B^2 / 2V$$

where *R* is the radius of curvature of the ion tube, *B* is the magnetic flux density (not *H*, the magnetic field intensity—the magnet's flux density divided by the permeability, *B*/μ), and *V* is accelerating potential of the ions.

quadrupole or **transmission-quadrupole** (single-letter symbol **Q**) mass spectrometer: This instrument separates ions based on oscillations in an electric field (the quadrupole field) created with the use of radio frequency (RF) and direct current (DC) voltages. For a given amplitude of a fixed ratio of RF (at a fixed frequency) to DC, ions of only a single *m/z* value will oscillate and pass from one end of a set of four poles positioned to form a hyperbolic cross section. All other ions are filtered out of the ion beam. This transmission of ions of a single integer *m/z* value is the reason why the term **mass filter** is sometimes used when describing this mass spectrometer. A mass spectrum is obtained by increasing the amplitude of the RF and DC while holding their ratio constant. These instruments operate between unit and 0.1 *m/z* resolution. The use of **transmission** as an adjective for this type of instrument is to avoid confusion with the other type of mass spectrometer that uses a quadrupole field for the separation of ions—the quadrupole ion-trap mass spectrometer (QIT-MS).

quadrupole ion-trap or **ion-trap** (symbol **QIT**) mass spectrometer (a.k.a. **Paul ion trap**): This type of mass spectrometer uses a quadrupole field to separate ions. The use of **quadrupole** as an adjective for this type of instrument is to avoid confusion with the other type of ion-trap mass spectrometer—the FTICR-MS). This instrument is also referred to as a *Quistor* (quadrupole ion storage) mass spectrometer.

There are two specific types of quadrupole ion-trap mass spectrometers: 1) instruments where the primary ionization takes place inside the place where ions are stored—the **internal ionization quadrupole ion trap** (I^2**QIT**); 2) instruments where the primary ionization takes place outside the place where ions are stored—**external ionization quadrupole ion trap** (**ExIonQIT**). LC/MS requires the use of an ExIonQIT-MS. GC/MS can be carried out with an ExIonQIT-MS or an I^2QIT-MS.

The QIT mass spectrometer uses a three-dimension quadrupole field for ion separation. Ions of a range of *m/z* values (based on the frequency of the RF voltage) are stored in a field that is created with a fixed-frequency RF applied to a cylinder ring electrode. End-caps held at ground, or subjected to various wave forms at different-frequency RF or DC voltages, are positioned on either side of the ring electrode but are electrically isolated. As the amplitude of this fixed-frequency RF is increased, the trajectories of ions of increasing *m/z* values will become sequentially unstable and move toward the two end-caps. Those ions reaching the end-cap (or caps) with holes to allow ions to pass to a detector are recorded. The increasing instability of the ions is observed as an increased amplitude of the orbital path of the ion.

The application of various wave forms to the end-caps can result in the selected storage of precursor ions for MS/MS or ions specific to an analyte with the exclusion of matrix ions that could result in reduced detection limits. Specific ion storage also allows for very low pressure chemical ionization (10^{-5} Torr for the reagent gas as opposed to 1 Torr in a beam-type instrument). Commercial models of QIT mass spectrometers operate at unit to 0.1 *m/z* resolution; however, a resolving power of 10^7 has been demonstrated (March, 1995).

time-of-flight (TOF) mass spectrometer: This type of instrument uses no external force to separate ions of different *m/z* values. Ions are accelerated into a flight tube at a few hundred to several thousand volts (30 kV). The ions will have differing velocities based on their mass. Lighter ions reach the end of the flight tube and are detected before the heavier ones. The equation that describes the ion separation in a TOF mass spectrometer is:

$$m/z = 2Vt^2/D^2$$

where *m* is the mass of the ion, *V* is the acceleration voltage of the ion, *D* is the length of the flight tube, and *t* is the time from which the ion is accelerated until it reaches the detector at the end of the flight tube. The inherent resolving power of the TOF-MS is not very great; however, it can be enhanced with techniques of orthogonal ion injection and the use of reflective ion mirrors (see **Terms Associated with Time-of-Flight Mass Spectrometers**). The TOF-MS was developed in the late 1950s, was very popular in the 1960s and 1970s, fell into almost total disuse in the 1980s (except in Russia), and saw a resurgence in the 1990s because of MALDI and rapid data acquisition techniques.

TERMS ASSOCIATED WITH DOUBLE-FOCUSING MASS SPECTROMETERS

decade: This term is used in conjunction with magnetic-sector and double-focusing mass spectrometers to describe an order-of-magnitude change in the *m/z* data acquisition range. Magnetic-sector mass spectrometers that obtain a spectrum using a fixed-field magnet and vary the accelerating voltage (no longer commercially produced) should only reduce the initial voltage by an order of magnitude. Lower values will result in decreased ion energy that will cause decreased sensitivity at the high end of the spectrum. A change in the magnetic field strength allows a different *m/z* decade range to be acquired for the same decade reduction in voltage. If an acquisition range of *m/z* 30–550 is required, then two separate scans have to be carried out using two different magnetic field strengths: one from *m/z* 30–300 and the other over the range of *m/z* 60–600. Although the first would require the full-decade scan time, the second would only require the fraction that is needed to acquire data from *m/z* 301–550; however, some overlap should be used to assure that a complete spectrum is obtained.

In modern instruments, the magnet can be scanned linearly (incrementing a DAC; only done by JEOL) or exponentially:

$$M = M(0)e^{kt}$$

where $M(0)$ is the starting *m/z* value, k is a constant, and M is the *m/z* value at the time t. More than a decade can be scanned, but the scan-time equation uses the time to scan a decade (20–200, 60–600, etc.) to define the scan speed for an exponential scan, which has the advantage that all peaks have a constant width in time regardless of *m/z* value. An exponential magnet downscan is also easy to implement in an analog fashion by discharging a capacitor. Linear magnet scans emphasize higher mass peaks. Linear scans can start at *m/z* 0, but *m/z* 0 can never be obtained with an exponential scan.

forward geometry: This term describes a double-focusing mass spectrometer in which the electric sector follows the ion source and precedes the magnetic sector. This instrument has an **EB** configuration.

reverse geometry: This term describes a double-focusing mass spectrometer in which the magnetic sector follows the ion source and precedes the electric sector. The electric sector is the last field that is experienced by the ions prior to their entering the detector or a collision cell. This instrument has a **BE** configuration.

field-free region: This term (abbreviated **FFR**) is not necessarily limited to sector-based mass spectrometers. It is any region of the mass spectrometer where the ions are not subjected to a field (electric, magnetic, quadrupole, etc.). Avoid using arcane terms such as "first FFR" or "second FFR." Use "…the field-free region traversed by the ion beam before entering the electric sector" or "…the field-free region between the electric and magnetic sectors."

MIKES and **IKES**:   These are abbreviations for **mass-analyzed ion-kinetic-energy spectrometry** and **ion-kinetic-energy spectrometry**, respectively. These techniques are used to obtain a product-ion mass spectrum of a metastable transition or product ion produced from a precursor ion in a collisionally activated dissociation. MIKES can only be accomplished in a reverse-geometry double-focusing mass spectrometer (Watson, 1997).

metastable ions: The electron ionization process produces three categories of molecular ions: 1) stable ions – those ions that remain intact for the 100 μs or longer needed to reach the detector; 2) unstable ions – those ions that decompose within $<10^{-7}$ s and are detected as fragment ions; 3) metastable ions – those ions that decompose outside of the ion source within 1–100 μs after their formation. Metastable ions are accelerated from the ion source as one species (precursor ions), undergo decomposition as a consequence of energy deposited during ionization, and are detected as product ions.

TERMS ASSOCIATED WITH TIME-OF-FLIGHT MASS SPECTROMETERS

coaxial reflectron: This term refers to a TOF instrument in which the ion source, reflectron, and detector are on the same axis (in line with one another). Ions travel in a line from the ion source to the reflectron, back in the direction from which they entered the reflectron through the ion source, and into the detector, which is in front of the ion source.

curved-field reflectron: This type of device allows for the energy-focusing of ions over a broad *m/z* range to produce a complete mass spectrum from a single laser shot that induced post-source decay of a large molecular mass analyte (i.e., proteins).

draw-out pulse: (a.k.a. **push-out pulse**) This term is used in reference to the voltage used to remove the ions from the ion source in a TOF mass spectrometer. This process of ion removal is accomplished by the pulsing of the ion-extraction field.

delayed extraction: (originally known as **time-lag focusing**) The delayed extraction technique is the time delay between the ionization pulse and the draw-out pulse. This technique is used to improve mass resolving power.

ion-gate pulse: This technique is the application of an electrical pulse between either the ion source and the **flight tube** (a.k.a. **field-free region** or **drift-tube region**) of the TOF mass spectrometer or the flight tube and detector to allow ions of only a narrow *m/z* range to pass.

gridless reflectron: Because there is scattering of ions after passing through grids (and actual wire mesh) that make up the ion mirror in a TOF-MS (which can result in ion losses ~10%), the gridless reflectron was developed. The gridless reflectron uses a series of rings with varying potential applied to create the mirror rather than the grids. Tight focusing of the ion beam close to the reflectron axis is required because divergence of ion trajectories for off-axis ions can result.

orthogonal extraction: Ions are pushed into the extraction region. A pulse is applied at a 90° angle to the direction of the ion flow to push the ions into the flight tube of the TOF-MS.

post-source decay: Post-source decay (PSD) is the fragmentation of ions (once they have been fully accelerated) in the drift tube prior to entering the reflectron, if one is present. Although post-source decay is a term often used in conjunction with time-of-flight mass spectrometry, the term does not depend on the style or pressure of a certain analyzer—it is an ion property.

reflectron: The reflectron (referred to as an ion mirror in its simplest form) is used to change the direction in which the ions are traveling (reflecting field) and to energy-focus ions (retarding field) for improved resolution. A single-stage reflectron has a single-retarding reflecting field, whereas a dual-stage reflectron has two linear-retarding voltage (constant field) regions that are separated by an additional grid. A dual-stage reflectron allows for smaller flight-tube lengths compared to single-stage reflectrons.

The following two terms are associated with MALDI MS. Although any type of *m/z* analyzer can be used with MALDI, the TOF-MS is most often employed.

fluence: This term is associated with laser-induced ionization such as in MALDI. This term is borrowed from physics, but it has a somewhat different definition in mass spectrometry. In mass spectrometry, fluence relates to the energy of the laser, the size of the impact area, the cross-sectional area of the sample, and the time of the laser **pulse**. In physics, energy fluence (Ψ) is expressed as a function of time (i.e., the energy fluence rate or energy flux density).

ablation: As used with reference to mass spectrometry, this term refers to the process of removing particles from the surface with a laser. In mass spectrometry, the removed (laser-ablated) molecules are ionized by charge transfer (one theory for what takes place in MALDI), photoionization, or through energy from a second laser. In MALDI, it has been theorized that the analyte molecules entrained in the matrix **plume**, which is undergoing a supersonic expansion, are ionized by the matrix molecules that have been ionized while still in the solid phase in the matrix.

Ablation is one of a series of words taken by mass spectrometry to describe a phenomenon that happens in the technique that has no single describing term. As defined in the dictionary, ablation means: 1) surgical excision or amputation of a body part or tissue; 2) the erosive processes by which a glacier is reduced; 3) in aerospace, it is the dissipation of heat generated by atmospheric friction, especially in the atmospheric re-entry of a spacecraft or missile by means of a melting heat shield. From Latin *ablātus*, past participle of *auferre*, to carry.

USE OF ABBREVIATIONS

Use of abbreviations in books, reports, and articles is a necessity in order to have an easy-to-read document and to have a document that is not excessive in length. Some reports on mass spectrometry terminology list a series of abbreviations that can be used without definition. All abbreviations should be defined before their use. In lengthy documents, an abbreviation should be defined multiple times in order to produce clarity. Nothing is more frustrating than to follow an index to a page in the middle of a 500-page book that contains several esoteric abbreviations that were defined in several isolated places in previous chapters. In such cases, it may be advisable to include a table of abbreviations. In a recent book on mass spectrometry, such a table had more than 100 items, which allowed for a reasonable size book and facilitated its use as a reference (de Hoffmann, 1996). The following is a guide to some abbreviation suggestions.

A great deal of care must be taken in the use of abbreviations to describe instruments such as a **GC-MS** and analytical techniques such as **GC**, **MS**, and **GC/MS**. This care is especially true with respect to "hyphenated instruments" and with techniques that use hyphenated instruments. **GC** or **LC** is almost always used to refer to an instrument—a **gas chromatograph** is a **GC** and a **liquid chromatograph** is an **LC**. However, **GC** can also be used to refer to the technique of **gas chromatography**; **LC** can also be used to refer to the technique of **liquid chromatography**.

Sometimes, **MS** is used to refer to an instrument—a **mass spectrometer** (*e.g., the time-of-flight MS can have good mass accuracy*); and other times, it is used to refer to a piece of data—a **mass spectrum** (*e.g., the MS of cocaine is quite distinctive*); or **MS** can also be used to refer to the technique of **mass spectrometry** (*e.g., MS is one of the most used analytical techniques*). Care must be taken in the use of these abbreviations to clarify what is meant. It would be considered bad form to use one of these abbreviations to mean two or more different things in the same document. Because the use of the words **mass spectrometer**, **mass spectrum**, and **mass spectrometry** make for long documents and tedious reading, one of the three is often abbreviated. The word **spectrum** can be substituted for mass spectrum. The use of the **MS** abbreviation for both mass spectrometer and mass spectrometry in the same document can be understandable by those with English as a first language; but for those with some other first language, this convention can be confusing. Abbreviations should be defined the first time they are used, and then used as often as required for clarity when it is not possible to include a table of abbreviations.

Some references have been made to "hyphenated techniques." Hyphenated techniques should mean techniques that use hyphenated instruments. The slash (/) is used to separate a combined technique such as gas chromatography/mass spectrometry (GC/MS) because the result of the chromatography is being analyzed by the mass spectrometer. Although the *Current IUPAC Recommendations* specifically states that GC/MS and LC/MS can be used without definition for either the combined techniques or the combined instruments, this practice can create confusion and should be avoided. All abbreviations should be defined.

GC/MS is used as the abbreviation for the technique of **gas chromatography/mass spectrometry**. **GC-MS** is used as the abbreviation for a **gas chromatograph-mass spectrometer** because the instrument combines a mass spectrometer with a gas chromatograph. Therefore, **LC/MS** refers to **liquid chromatography/mass**

spectrometry; **LC-MS** refers to a **liquid chromatograph-mass spectrometer**. **CE/MS** refers to **capillary electrophoresis/mass spectrometry**; **CE-MS** refers to a **capillary electrophoresis-mass spectrometer**.

A single analytical technique is abbreviated without a hyphen or a slash (/). Therefore, when using **MS** as an abbreviation for mass spectrometry, the correct abbreviation for time-of-flight mass spectrometry would be **TOFMS**; however, when using **MS** as the abbreviation for a mass spectrometer, the correct abbreviation for a time-of-flight mass spectrometer would be **TOF-MS** (the two are separated by a hyphen).

When the ionization method is used in the description of the mass spectrometer, the ionization method and the instrument type are separated by a space (i.e., **EI TOF-MS** for an electron ionization time-of-flight mass spectrometer). A space following the ionization type would also be correct for inclusion with the ion-separation technique (i.e., **EI TOFMS** for electron ionization time-of-flight mass spectrometry).

The above stated use of the hyphen and slash is supported by the Information to Authors section in recent issues of *Rapid Communications in Mass Spectrometry.* Other mass spectrometry journals do not specifically address this issue in their author-instruction sections.

The technique of MS/MS involves more than a single *m/z* analyzer, except in the case of QIT and FTICR mass spectrometers. Multiple *m/z* analyzer instruments are sometimes described with an abbreviated notation that uses the single-letter symbols for the instrument type described above. The lower case **q** is a single-letter symbol used to describe the location of an RF-only quadrupole collision cell. Some examples are: **QqTOF** for a transmission-quadrupole *m/z* analyzer, followed by a quadrupole collision cell, followed by a time-of-flight *m/z* analyzer; and a **QqQ** as the abbreviation for a triple-quadrupole mass spectrometer.

When to use alpha, α, alpha-, or α-, etc.
An important nomenclature convention is the use of spelled-out Greek words as opposed to the Greek letters (e.g., alpha vs α). When the nomenclature **alpha**, **beta**, and **sigma cleavage** was first proposed (Budzikiewicz, 1964), the authors incorrectly elected to use chemical-name conventions. Just as β-naphthol was written with a hyphen to indicate that the functionality is at the beta position of the molecule, α-cleavage was written with a hyphen. This style placed the index reference for α-cleavage under **C**. Subsequent authors (Watson, 1997; McLafferty, 1993) refer to the breaking of the alpha bond as alpha cleavage in the index and place the reference under **A**, which is more correct as an ease-of-use feature. In the case of alpha cleavage, alpha is an adjective that modifies cleavage, which is a noun; therefore, alpha cleavage is not a hyphenated term. These latter authors, who indexed alpha cleavage, use the term α-**cleavage** incorrectly in the text of their books. If the symbol is used, there is no need for a hyphen because the symbol is used as a shorthand form of the adjective (alpha) that modifies the noun (cleavage). The α does not specify the location of a functionality in a molecule or an ion. The correct term is α **cleavage**. However, when α cleavage is used to describe an event (e.g., an α-cleavage reaction), then a hyphen is used because the term α-**cleavage** is an adjective that describes reaction, which is a noun.

From the early mass spectrometry literature through the present, the words and the Greek letters were used for alpha, beta, and sigma. The first edition of *The ASC Style Guide* (Dodd, 1986) states: "Greek letters, not the spelled-out forms, are used in chemical and drug names."

Unfortunately, the examples that are given use the hyphen with the Greek letter and omit the hyphen when the spelled-out name is used (e.g., β-naphthol vs beta naphthol). In this example, a hyphen should have been used regardless of the chosen form because beta is a prefix, NOT an adjective.

According to the second edition of *The ASC Style Guide* (Dodd, 1997), the Greek letter is the preferred style in ALL cases: "Use Greek letters, not the spelled-out words, for chemical and physical terms."

The examples given in this case are γ radiation versus gamma radiation and β particle versus beta particle. This particular style change was probably a result of the desire for less print in articles, but presents a dilemma for mass spectrometry. For the sake of clarity, the spelled-out word should be used where needed. If a Greek letter (e.g., α, β, or σ) is used to describe a cleavage, then no hyphen is necessary when the letter is meant as a shorthand notation for the word used as an adjective that modifies cleavage.

Remember, a γ-hydrogen (the γ is a prefix) shift-induced cleavage requires the use of a hyphen, whereas no hyphen is required to describe the γ hydrogen (the γ is a shorthand notation for gamma, an adjective) that has shifted to bring about the cleavage.

COMPONENTS OF A MEASUREMENT

There are four components to a measurement: signal strength, baseline offset (offset from zero), background (signal strength due to background), and noise (reproducibility of the signal from one sample to the next).

signal strength	a measured response of the system to a precise amount of sample.
baseline offset	the difference between zero of the system and the lowest level that a signal can be observed when no sample is present.
background	the level and variation of the signal when ionization is on, but no sample is present.
noise	the variation of signal strength when a sample is present from one measurement to the next.

At the lower levels of detection, the noise and the background will be close enough to the same value to be considered equal; therefore, the detection limit of a system can be defined as the ratio of the signal strength produced by the sample to the variation from one measurement to the next (signal-to-noise) or the ratio of the signal strength produced by the sample to that of the background for a measurement (signal-to-background). If signal-to-background is used as the measurement of sensitivity, then the background signal must be measured in the same sampling interval as the sample signal and should be measured as near as possible to the sample signal. Whether sensitivity is determined as an S/N or S/B, the value for the zero offset must be taken into account.

The **detection limit** is the minimum amount of sample that can produce a signal that allows the determination of the presence of the sample, and is expressed as an S/N or S/B value relative to an amount (e.g., 5 pg of hexachlorobenzene will produce a mass chromatographic peak that has an S/N value greater than 10). Detection limits usually assume that only the specified analyte is producing the signal. In some cases, a detection limit may specify a criteria for confirming the presence of the analyte (e.g., in the above example, it may be stated that the sample must produce a library searchable spectrum).

The **limit of quantitation** is the smallest amount that can be measured to within a specified precision.

The precision of a numerical result is the degree of agreement between the result and other values obtained under the same conditions. The percent precision is 100 divided by the smallest numeric value that can be measured under a given set of conditions (e.g., a 50% precision will allow a value as small as 2 to be measured, a 20% precision will allow a number of no less than 5 to be measured, and a 5% precision will allow the measurement of a number of no less than 20).

The **limit of quantitation** becomes the smallest number that can be measured times the noise. If the number of units of noise in a measurement is 5 and a 5% precision is required for the limit of quantitation, then the smallest measurement allowed is 100 units. It should be remembered that the level of background and the level of noise are the same at the lowest signal level.

FORMULAS AND EQUATIONS

If a formula or an equation contains mathematical operators (+, −, =, ×) that represent an arithmetic function where the + and − does not relate to the sign of a number (or symbol) or the result of the product of a set of numbers (or elements), then the operator should be preceded and followed by a space. No spaces precede or follow the slash (/) when used as a mathematical operator. Parentheses and brackets should be presented without preceding or trailing spaces, except when using the arithmetic operators.

Examples:

$E = mc^2$

NOT $E=mc^2$ or $E = m\ c^2$

$A + B = C$

NOT $A+B=C$ or $A+B = C$

$(a/b)(c/d)$

NOT $(a\ /\ b)(c\ /\ d)$

$1 + [2A − B] = [C + D]$

NOT $1+[2A−B]=[C+D]$ or $1 + [\ 2A − B\ ] = [\ C + D\ ]$

$PV = nRT$

NOT $PV=nRT$

To be consistent with IUPAC recommendations, the symbols in equations should be in italics, but not the operators.

TYPES OF ELEMENTS AND ELECTRONS

For the purpose of describing the various types of elements with respect to the isotopic ratios that are used in the determination of the elemental composition of an ion, the common elements found in organic mass spectrometry are referred to as:

X elements   those elements that exist as only a single naturally occurring stable isotope (H, F, P, I). Because of the low natural abundance of deuterium compared to hydrogen, hydrogen is considered an X element.

X+1 elements   those elements that exist as only two naturally occurring stable isotopes, and that are 1 integer u different (C and N).

X+2 elements   those elements that exist as more than a single naturally occurring stable isotope, and two of the isotopes differ in integer value by 2 u (Cl, Br, S, Si, and O).

It should be noted that the letter **A** rather than **X** has been used to describe these different types of elements. However, because **X** is a universally accepted symbol for an unspecified number, and unspecified ions or peaks can occur with various values, the use of **X** is more descriptive and therefore is recommended.

An organic compound, molecule, ion, and/or radical has three types of electrons:

n- or **nonbonding electrons**, which are not associated with the attachment of one atom to another.

π- or π-**bond electrons**, which are found in a chemical bond formed by the overlap of two adjacent *p*-orbitals.

σ- or σ-**bond electrons**, which are associated with bonds formed by the overlap of an *sp*, *sp^2*, or *sp^3* hybrid orbital, and another such hybrid orbital or an *s*-orbital.

TERMS ASSOCIATED WITH COMPUTERIZED SPECTRAL MATCHING

Library Search

There are a number of data-system-specific terms that are used with various mass spectral library search programs. They can be reviewed in the documentation used with each manufacturer's data system. The following terms have appeared in the literature and are noteworthy.

fit: This term is specific to the INCOS library search system distributed by Finnigan and Varian. Peaks that are in the sample spectrum and not in the library spectrum are disregarded in evaluation of the match. This type of comparison has the advantage of being able to identify multiple compounds represented by a single spectrum.

match factor: This term is used to describe the quality of the match between the sample spectrum and the spectrum found by the **NIST Mass Spectral Search Program for Windows** (**NIST MSS Program**). It has the same meaning as **purity** in the INCOS search. It is based on the dot-product result of a point-by-point comparison of the two spectra.

purity: This INCOS-specific term (as distributed by Varian and Finnigan) is used instead of **similarity index**. It takes into account all peaks in the sample and library spectra to make a determination of how close the two spectra match. The purity is calculated by using mass-weighted intensities in conjunction with a linear-algebra dot product. For more information on this term, consult the data-system manuals from Finnigan or Varian.

probability: This term has two different meanings. The first meaning is associated with the **Probability Based Matching** (**PBM**) algorithm as implemented by Agilent Technologies, Palisade Corporation, Teknivent, ThruPut Software, and others. This term is supposed to be a value that represents the probability that the "unknown is correctly identified as the reference." However, the HP G1034C MS ChemStation Software manual (ca. 1993) states, "Values less than 50 mean that substantial differences exist between the unknown and reference, and the match should be regarded with suspicion." This wording sounds like a comparison of the two spectra is being made, which would mean that the **probability** term is being used in the same way as the **purity** term of INCOS and the **similarity index** term of NIST (a probability that the two spectra are from the same compound based on a direct mass-and-intensity comparison).

The second definition of the term **probability**, as used in the NIST MSS Program, describes the likelihood that the mass spectrum of the unknown and the mass spectrum retrieved from the database are of the same compound. The probability is based on all the matches found during the search. In the case of a search of a submitted spectrum of m-xylene, there would be matching spectra of all three isomers of xylene found in the library. All three would have very high match factors when compared with the submitted spectrum, and would mean that the probability of any one of the three library spectra being of the same compound as that which produced the submitted spectrum would be divided among the three and would therefore be low. The NIST MSS Program is the only program that offers this type of comparison.

quality: This term is used in the Agilent Technologies implementation of the **PBM** mass spectral search algorithm to describe how closely the submitted spectrum and the library spectrum match. It is not clear from the HP documentation how the word **quality** used in this case differs from the **probability** value reported for a match.

Rfit (reverse fit): **Rfit** is another INCOS-specific term. It means that the evaluation is based on the peaks in the submitted spectrum being in the library spectrum. The library spectrum can have additional peaks without affecting the value of the similarity.

reverse match factor: This term is unique to the NIST MSS Program. It is a **match factor** that is calculated by ignoring any peaks in the sample spectrum that are not in the library spectrum. These peaks are considered to be due to impurities. This term is similar in nature to the **fit** definition in the INCOS search and the **reverse search** value used in the NIST Mass Spectral Search Program for DOS.

reverse search: This term can also have two different meanings. The first meaning pertains to a sequential search of a data file of mass spectra to see if it contains a spectrum of a specific compound. This type of search is used in the analysis of GC/MS or LC/MS data for target compounds. The reverse search is the basis for many automatic search and quantitation programs. A retention-time window in the data file is searched to see if a spectrum is present that matches the spectrum of the target analyte.

The second definition of the term **reverse search**, as used in the NIST MSS Program, means the "hit list" is presented in decreasing order of the **reverse match factors** as opposed to decreasing order of **match factors**. This process treats the extraneous peaks in the submitted spectrum as impurities. The **reverse search** values in the NIST MSS Program are the same as the **fit** values used to describe the results of the INCOS library search provided by Finnigan and Varian.

similarity index (SI): This term has been used by a number of different mass spectral library search programs. The **similarity index** is a comparison of masses and relative intensities. SI was used with the NIST Mass Spectral Search Program for DOS, but abandoned in the Windows version. There is another term that has also been used with a similar meaning—**dissimilarity index (DI)**. Both of these terms are used to describe how similar the library spectrum is to the submitted spectrum. Neither similarity index nor dissimilarity index is used with current commercial data systems.

Search Algorithms

There are primarily two search algorithms in use in most commercially available data systems: 1) the Probability Based Matching (PBM) algorithm (McLafferty, 1974); and 2) the dot-product algorithm (a.k.a. INCOS (Sokolow, 1978), which is associated with a Finnigan Corporation-specific data system). A number of enhancements and variations to the INCOS algorithm have been made and are distributed by the National Institute of Standards and Technology (NIST), Micromass, Finnigan, Varian, Agilent Technologies (on their GC/MS and LC/MS products), Perkin-Elmer (on their GC/MS product only), and others. The PBM algorithm is used by Waters (in their LC/MS system); and Teknivent, Shimadzu, and ThruPut in their GC/MS systems. The PBM algorithm is distributed by Agilent Technologies, along with the NIST version of the INCOS algorithm on their GC/MS products only. The variations in the implementation of the INCOS algorithm primarily have to do with the presearch and spectral condensations.

The presearch is a system by which possible candidates for a match to the unknown spectrum are selected from the mass spectral database. This presearch is carried out by using a subset of the unknown spectrum and the database spectra (one or more types of spectral condensations). Some presearch algorithms involve only a reduced sample spectrum (8 most intense peaks) and less condensed database spectra (16 most intense peaks). Other algorithms employ multiple criteria. The presearch is an important area of consideration. If the spectrum of the sample compound is excluded in the presearch, then the results are much like "throwing the baby out with the bath water." There have been other algorithms that existed in the formative years of mass spectral data systems. Some are included in the following brief descriptions.

PBM: The Probability Based Matching (PBM) algorithm uses a **weighting** and **reverse search**. The **weighting** is used to determine the importance of peaks based on masses and relative intensities. The weighting is used in the presearch. If the sample spectrum and the library spectrum of the same compound do not have the same base peak, then it is very likely that the library spectrum of the matching compound will be excluded. The probability that particular abundances will occur follows a log-normal distribution. The probability of most mass values also varies in a predictable manner. Because the larger molecular fragments tend to decompose to give smaller fragments, the probability of higher m/z values decreases by a factor of two approximately every 130 m/z units. This weighting system is used for indexing the "Important Peak Index of the Registry of Mass Spectral Data." The second feature of PBM, **reverse searching**, treats peaks in the submitted spectrum and not in the library spectrum as if they are from another compound. This process is valuable in identification of the spectrum of more than one compound. The PBM algorithm compares the submitted spectrum against the "Important Peak Index." Spectra found by this comparison are evaluated against the spectra in the entire database or a condensed version of it. Because of the way that the "Important Peak Index" is developed, the PBM search algorithm is limited in its performance when small databases of target compounds are considered.

INCOS: The INCOS library search system uses a presearch of matching the 8 most intense peaks in the submitted spectrum with a set of the 16 most intense peaks in the library spectrum. A search is performed against the 50 most chemically significant peaks in the spectra that are retrieved by the presearch. This search uses a weighting of the intensity so that peaks of higher *m/z* values are more significant than peaks of lower values. The comparison is done relative to adjusted abundances, which are the square root of the observed peak intensity times the *m/z* value. The results are reported for **purity**, **fit**, and **Rfit** (see above). Unfortunately, the condensation algorithm to produce the condensed database used by Finnigan and Varian is not very good with respect to which peaks are discarded. A condensation of a full-spectrum database (such as the Full.d file produced for the Agilent Technologies system) using **MassTransit™** (Palisade Corporation) produces a far-superior condensed database because MassTransit uses the same **weighting** described with the PBM algorithm.

NIST: The library search system algorithm of the NIST MSS Program is similar to that of INCOS, except that the presearch and the main search are against a set of weighted intensities. The user has control over a number of different factors that can affect this search, such as whether or not **neutral loss** logic is used. This search program is offered by a number of different manufacturers in addition to their own routines.

N most int. Peaks: This algorithm is a comparison of a set of most intense peaks in the spectrum. This comparison is usually carried out against a set of eight peaks. This algorithm requires that the spectrum that corresponds to the submitted spectrum of the sample compound is in the mass spectral reference library. This algorithm has the advantage of being able to program easily, and the amount of disk storage space is limited.

N most sig. Peaks: This algorithm is a comparison of the most intense peaks after a weighting of intensities. This algorithm also requires that the spectrum that corresponds to the submitted spectrum of the sample compound is in the mass spectral reference library. This algorithm has the same advantages as the **N most int. Peaks** algorithm.

Biemann search: This algorithm is a comparison of the two most intense peaks within each window of 14 *m/z* units, starting the divisions at *m/z* = 6. This algorithm has the same limitations and advantages as the **N most int. Peaks** and **N most sig. Peaks** algorithms (Hertz, 1971).

neutral loss: This algorithm requires that the matching spectra must have peaks that correspond to neutral loss peaks in the submitted spectrum. This type of search requires that the nominal mass of the compound of the submitted spectrum is known. The neutral loss search usually is used in combination with other algorithms and requires a database of neutral losses.

RTI & spectrum: The retention-time index (RTI) and spectrum-match algorithm is most often used with target-compound analyses. This type of search compares all the spectra that are in a given retention-time window and determines if there is a reconstructed chromatographic peak that contains spectra that have the mass spectral peaks found in the reference spectrum. This algorithm uses a **reverse search** because it does not consider peaks in the sample spectrum that are not in the library spectrum.

ION DETECTION

The ions that are *m/z*-analyzed by a mass spectrometer are detected with several different devices:

dynode: (**high-energy** or **conversion**):   A dynode is an element in an electron tube (**electron multiplier**: **EM**) whose primary function is to provide secondary emission of electrons.   Secondary emission occurs when an electron that moves at sufficiently high velocity strikes the dynode.   Positive and negative ions, as well as fast moving neutrals, can also be used to produce an emission of secondary electrons.   In a transmission-quadrupole mass spectrometer (TQ-MS) and a quadrupole ion-trap mass spectrometer (QIT-MS), higher-mass ions strike the detector with less velocity than lower-mass ions.   This variation in velocity of ions can result in mass discrimination.   A **high-energy dynode** is used to accelerate the ions so that the secondary emission caused by the impact of the ion on the electron multiplier's first secondary emission dynode is more equal regardless of its mass.   Similar actions are taken on high-mass MALDI ions mass-resolved in a TOF-MS.   The addition of a high-energy dynode has been used to produce the **post-acceleration detector** (**PAD**).

A **conversion dynode** is used to convert negative ions to positive ions in the TQ-MS and QIT-MS.   The first dynode of an EM has a negative bias to attract positive ions.   Due to the low-ion acceleration in a TQ-MS (10–15 V), the negative ions are repelled.   The use of a conversion dynode is not necessary in a magnetic-sector mass spectrometer because the ions strike the detector with 3–10 kV of acceleration.

electron multiplier: (**continuous-** and **discrete-dynode EM**): An **electron multiplier** (**EM**) is a device used for current amplification through the secondary emission of electrons.   Typically, the electron multiplier will produce 10^5 electrons ion^{-1} (a gain of 10^5).   A **discrete-dynode EM** has a series of secondary emitters (**dynodes**).   When an ion strikes the first dynode with sufficient velocity to produce the emission of a number of electrons (the emitting surfaces were constructed of a copper-beryllium material in earlier vintages), the electrons are attracted by the positively biased second dynode, where even more electrons are produced.   This process is repeated through several generations of secondary emissions. It is not uncommon for a discrete-dynode EM to have 10–20 stages.   The typical operation voltages for this EM is 3–5 kV.   The **continuous-dynode EM** uses a continuous surface of lead-doped glass that produces the same effect. The typical operating voltage of a continuous-dynode EM is <2 kV.   Currently, there are also discrete-dynode electron multipliers that use the lead-doped glass and operate at a lower potential.

microchannel plate: (a.k.a. **channel electron multiplier**: **CEM**) The CEM is an array of 10^4–10^7 **continuous-dynode** electron multipliers (10–100 µm dia.), side by side, each acting as an independent emitter to form a spatially resolved array detector. This type of detector is often used in conjunction with phosphor plates and a photodiode array to replace photographic plates in focal-plane mass spectrometers (mass spectrographs). They are also commonly found in TOF mass spectrometers.

Daly detector: In this detector, ions strike a dynode to produce secondary electrons that are directed toward a phosphor screen. When the electrons impact on the screen, photons are emitted and amplified with a photomultiplier tube. Unlike the EM or CEM, the amplification device is outside the vacuum system of the mass spectrometer. The phosphor screen experiences far less wear per ion impact than either the EM or CEM; therefore, this type of detector has a much longer life (several years as compared to 12–24 months or less for the EM and CEM).

Faraday cup: Positive ions arriving at a metal surface are neutralized by electrons that have passed through a high-ohm resistor. The ion current is calculated from the voltage drop across the resistor. A guard tip is required to prevent the loss of secondary electrons emitted by impacting ions. The measured and amplified ion current is directly proportional to the number of ions and to the number of charges per ion. The Faraday cup is an "absolute detector" and can be used to calibrate an electron multiplier. These detectors are free from mass discrimination because of a lack of dependence on ion velocity. This detector has a very long time delay that is inherent in its amplification system. It is used primarily for highly accurate measurements of slowly changing ion currents, such as the case with isotope-ratio mass spectrometry. However, even with this response-time limitation, a detector has been proposed (based on the Faraday cup principle) that has an electronic signal amplification system fast enough to mass- and time-resolve MALDI TOFMS signals.

---

* Bahr, U; Röhling, U; Lautz, C; Strupat, K; Schürenberg, M; Hillenkamp, F "A Charge Detector for TOF-MS of High-Mass Ions Produced by MALDI," *Int. J. Mass Spectrom. Ion Proc.* **1996**, *153*, 9–22.

REFERENCED CITATIONS IN DOCUMENTS
(Reports, Journal Articles, Chapters, etc.)

Different journals and publishers have different styles for referenced articles and book chapters. They are usually contained in the Instructions to the Author section of journals. The names of the authors, journal name, year, volume number, and page number or page-number range are always required. The inclusion of the title of the referenced article or chapter is somewhat contentious. Although not the case with most modern biotechnology journals, many chemistry and mass spectrometry journals not only do not require titles in cited articles, but they also do not allow for them.

The overwhelming volume of information that is available in mass spectrometry (as well as all scientific areas) today requires a high degree of efficiency in researching a topic. Most users of mass spectrometry do not have sufficient time to review title listings or abstracts of articles in their field, let alone track every referenced citation in a research article. However, when a citation does not contain the title of an article, the determination of the value of one of these references often requires that the original article or chapter be consulted. The volume of information and the time demand on the researcher make burdensome the need to consult the original citation in the absence of its potential value. The lack of titles in the citations in review articles, such as those that appear from time to time in the Accounts and Perspective section of the *Journal of the American Society for Mass Spectrometry*, *Mass Spectrometry Reviews*, the Special Features section of the *Journal of Mass Spectrometry*, and the A-Pages of *Analytical Chemistry*, detracts from the utility of these articles.

These and other reasons have compelled many scientific research funding agencies such as the U.S. National Institutes of Health to require the inclusion of titles with all citations in grant proposals. *Mass Spectrometry Reviews* now requires the inclusion of titles, and the *Journal of the American Society for Mass Spectrometry* (Gross, Sparkman, 1998), the *Journal of Mass Spectrometry*, and the *International Journal of Mass Spectrometry and Ion Processes* have just made editorial policy changes to allow (but not require [the policy is that if any cited reference includes a title, then all the citations within the article must include titles] consistency) for the inclusion of titles in cited references. The title of a scientific article is really a mini-abstract. When an article that contains titles in its references is reviewed, it is much easier to determine the potential value of retrieving one of these references.

The primary objection to the inclusion of titles in referenced citations is that many authors have extensive bibliographies without titles stored in their databases. The reason stated by many journals was the amount of space required; however, with the extensive amount of "white space" in most modern journals, this objection is without any reasonable merit.

It is hoped that future contributors to the scientific literature will see the value that titles add to references and will include them in citations, and that this practice becomes commonplace in submitted articles.

Note: The source for journal title abbreviations is *The ACS Style Guide: A Manual for Authors and Editors*, 2nd ed. (Dodd, 1997).

REFERENCES

1. Abramson, FP "Chemical Reaction Interface Mass Spectrometry," *MS Reviews* **1994**, *13*, 341–356.

2. Adams, J "Letter-To-Editor," *J. Am. Soc. Mass Spectrom.* **1992**, *3*, 473.

3. Aston, FW *Isotopes*; Edward Arnold: London, 1924.

4. Badman, ER; Wells, JM; Bui, HA; Cooks, RG "Fourier Transform Detection in a Cylindrical Quadrupole Ion Trap," *Anal. Chem.* **1998**, *70*, 3545–3547.

5. Barber, M; Bordoli, RS; Sedgwick, RD; Tyler, AN "Fast Atom Bombardment of Solids (F.A.B.): A New Ion Source for Mass Spectrometry," *J. Chem. Soc. Chem. Commun.* **1981**, 325–327.

6. Barnett, EF; Tandler, WSW; Turner, WR "Quadrupole Mass Filter with Fringe Field Penetrating Structure," U.S. Patent 3560,734, 1971.

7. Beckey, HD *Principles of Field Ionization and Field Desorption Mass Spectrometry*; Pergamon: New York, 1977.

8. Beckey, HD Field Ionization Mass Spectrometry. In *Advances in Mass Spectrometry*, Vol. 2; Elliott, RM, Ed.; Pergamon: Oxford, U.K., 1963; p 1.

9. Beynon, JH; Zerbi, G "IUPAC Recommendations on Symbolism and Nomenclature for Mass Spectrometry," *Organic Mass Spectrom.* **1977**, *12*(3), 115–118.

10. Beynon, JH "Report of Nomenclature Committee Workshop," *Proceedings of the 29th ASMS Conference on Mass Spectrometry and Allied Topics*, Minneapolis, MN, May 24–29, 1981; pp 797–815.

11. Biemann, K *Mass Spectrometry: Organic Chemical Applications*, Vol. 1; McGraw-Hill: New York, 1962. (reprinted as *ASMS Publications Classic Works in Mass Spectrometry*; American Society for Mass Spectrometry: Santa Fe, NM, 1998)

12. Blakely, CR; Vestal, ML "Thermospray Interface for Liquid Chromatography/Mass Spectrometry," *Anal. Chem.* **1983**, *55*, 750–754.

13. Brubaker, WM Quadrupole MS. In *Advances in Mass Spectrometry,* Vol. 4; Kendrick, E, Ed.; Institute of Petroleum: London, 1968; p 293.

14. Bruins, AP; Covey, TR; Henion, JD "Ionspray Interface for Combined Liquid Chromatography/Atmospheric Pressure Ionization Mass Spectrometry," *Anal. Chem.* **1987**, *59*, 2642–2646.

15. Budzikiewicz, H; Djerassi, C; Williams, DH *Interpretation of Mass Spectra of Organic Compounds*; Holden-Day: San Francisco, CA, 1964.

16. Bursey, MM; Hoffman, MK Mechanisms Studies of Fragmentation Pathways. In *Mass Spectrometry: Techniques and Applications*; Milne, GWA, Ed.; Wiley-Interscience: New York, 1971.

17. Bursey, MM "Editorial," *Mass Spectrom. Reviews* **1992**, *11*, 1, 2.

18. Bursey, MM "Editorial," *Mass Spectrom. Reviews* **1991**, *10*, 1, 2.

19. Caprioli, RM; Fan, T; Cottrell, JS "Continuous-flow Sample Probe for Fast Atom Bombardment Mass Spectrometry," *Anal. Chem.* **1986**, *58*(14), 2949–2953.

20. Cody, RB; Freiser, BS "Electron Impact Excitation of Ions from Organics: An Alternative to Collision Induced Dissociation," *Anal. Chem.* **1979**, *51*, 547–554.

21. Cooks, RG; Rockwood, AL "Letter-To-Editor," *Rapid Commun. Mass Spectrom.* **1991**, *5*(2), 93.

22. de Hoffmann, E; Charette, J; Stroobant, V Appendix 1: Nomenclature, Appendix 2: Abbreviations. *Mass Spectrometry: Principles and Applications*; Wiley: Chichester, U.K., 1996.

23. de Hoffmann, E "Tandem Mass Spectrometry: A Primer; Special Feature: Tutorial," *J. Mass Spectrom.* **1996**, *31*(2), 129–137.

24. Dempster, AJ "Positive Ray Analysis of Potassium, Calcium and Zinc," *Phys. Rev.* **1922**, *20*, 631–638.

25. Dodd, JS, Ed. *The ACS Style Guide*; American Chemical Society: Washington, DC, 1986.

26. Dodd, JS, Ed. *The ACS Style Guide: A Manual for Authors and Editors*, 2nd ed.; American Chemical Society: Washington, DC, 1997.

27. Dougherty, RC; Dalton, J; Biros, FJ "Negative Ionization of Chlorinated Insecticides," *Org. Mass Spectrom.* **1972**, *6*, 1171–1181.

28. Emmett, MR; White, FM; Hendrickson, CL; Shi, SD-H; Marshall, AG "Application of Micro-Electrospray Liquid Chromatography Techniques to FT-ICR MS to Enable High-Sensitivity Biological Analysis," *J. Am. Soc. Mass Spectrom.* **1997**, *9*, 333.

29. Emmett, MR; Caprioli, RM "Micro-Electrospray Mass Spectrometry: Ultra-High-Sensitivity Analysis of Peptides and Proteins," *J. Am. Soc. Mass Spectrom.* **1994**, *5*, 605.

30. Field, FH Chemical Ionization Mass Spectrometry. In *Mass Spectrometry* Vol. 5; Maccoll, A, Ed.; MTP Int. Rev. Sci. Butterworths: London, 1972; pp 133–138.

31. Ghaderi, S; Kulkarni, PS; Ledford, EB, Jr.; Wilkins, CL; Gross, ML "Chemical Ionization in Fourier Transform Mass Spectrometry," *Anal. Chem.* **1981**, *53*, 428–437.

32. Glish, G "Letter-To-Editor," *J. Am. Soc. Mass Spectrom.* **1992**, *2*, 349.

33. Gold, V "IUPAC Glossary of Terms Used in Physical Organic Chemistry," *Pure Appl. Chem.* **1983**, *55*, 1281 (specific definitions on pp 1296, 1297).

34. Gomer, R; Inghram, MG *J. Chem. Phys.* **1954**, *22*, 1279; *J. Am. Chem. Soc.* **1955**, *77*, 500.

35. Gross, ML "Editorial," *J. Am. Soc. Mass Spectrom.* **1994**, *5*, 57.

36. Gross, ML; Sparkman, OD "Editorial," *J. Am. Soc. Mass Spectrom.* **1998**, *9*, 451.

37. Hertz, HS; Hites, RA; Biemann, K "Identification of Mass Spectra by Computer Searching a File of Known Spectra," *Anal. Chem.* **1971**, *43*, 681–691.

38. Hillenkamp, F; Karas, M; Beavis, RC; Chait, BT "Matrix-Assisted Laser Desorption/Ionization Mass Spectrometry of Biopolymers," *Anal. Chem.* **1991**, *63*, 1193A–2003A.

39. Horning, EC; Horning, MG; Carroll, DI; Dzidic, I; Stillwell, RN "New Picogram Detection System Based on Mass Spectrometry with an External Ionization Source at Atmospheric Pressure," *Anal. Chem.* **1973**, *45*, 936–943.

40. Hunt, DF; Stafford, GC, Jr.; Crow, FW; Russell, JW "Pulsed Positive Negative CI," *Anal. Chem.* **1976**, *48*, 2098–2105.

41. Ito, Y; Takeuchi, T; Ishii, D; Goto, M "Direct Coupling of Micro High-performance Liquid Chromatography with Fast Atom Bombardment Mass Spectrometry," *J. Chromatogr.* **1985**, *346*, 161–166.

42. Karas, M; Bachmann, D; Bahr, U; Hillenkamp, F "MALDI," *Int. J. Mass Spectrom. Ion Proc.* **1987**, *78*, 53–68.

43. Karasek, FW; Clement, RE Gas Chromatography-Mass Spectrometry Glossary. *Basic Gas Chromatography-Mass Spectrometry Principles And Techniques*; Elsevier Scientific: Amsterdam, 1981.

44. Kiser, RW *Introduction to Mass Spectrometry and Its Applications*; Prentice-Hall: Englewood Cliffs, NJ, 1965; Introduction, p 1.

45. Kondrat, RW; Cooks, RG "Direct Analysis of Mixtures by Mass Spectrometry," *Anal. Chem.* **1978**, *50*, 81A–92A.

46. Lehmann, WD "Pictograms for Experimental Parameters in Mass Spectrometry," *J. Am. Soc. Mass Spectrom.* **1997**, *8*, 756–759.

47. March, RE; Todd, JFJ, Eds. *Practical Aspects of Ion Trap Mass Spectrometry*, Volume II *Ion Trap Instrumentation*; CRC: Boca Raton, FL, 1995.

48. McLafferty, FW *Interpretation of Mass Spectra*, 2nd ed.; Benjamin/Cummings: Redding, MA, 1973.

49. McLafferty, FW; Hertel, RH; Villwock, RD "Probability Based Matching of Mass Spectra," *Org. Mass Spectrom.* **1974**, *9*, 690–702.

50. McLafferty, FW; Tureček, F Glossary and Abbreviations. *Interpretation of Mass Spectra*, 4th ed.; University Science: Mill Valley, CA, 1993.

51. McLafferty, FW; Tureček, F *Interpretation of Mass Spectra*, 4th ed.; University Science: Mill Valley, CA, 1993; p 192.

52. Meng, CK; Mann, M; Fenn, JB "Of Protons or Proteins: A Beam's Beam," *Z. Phys.* **1988**, *D10,* 361–368.

53. Mills, I; Cvitaš, T; Homann, K; Kallay, N; Kuchitsu, K *Quantities, Units and Symbols in Physical Chemistry*, 2nd ed., International Union of Pure and Applied Physical Chemistry Division; Blackwell Science: London, 1993.

54. Munson, B; Field, FH "Chemical Ionization Mass Spectrometry," *J. Am. Chem. Soc.* **1966**, *88*, 2621–2630; Munson, B "CI-MS: 10 Years Later," *Anal. Chem.* **1977**, *49*, 772A–778A.

55. Nicholson AJC "Photochemical Decomposition of Aliphatic Methyl Ketones," *Trans. Faraday Soc.* **1954**, *50*, 1067–1073.

56. Olah, GA "Five-coordinated Carbon: Key to Electrophilic Reactions," *CHEMTECH* **1971**, *1*, 566.

57. Olah, GA "The General Concept of Carbocations Based on Differentiation of the Trivalent ('Classical') Carbenium Ions from Three-center Bound Penta- or Tetracoordinated ('Nonclassical') Carbonium Ions: The Role of Carbocations in Electrophilic Reactions," *J. Am. Chem. Soc.* **1972**, *94*, 808.

58. Price, P "Standard Definition of Terms Relating to Mass Spectrometry," *J. Am. Soc. Mass Spectrom.* **1991**, *2*, 336–348.

59. Radom, L; Bouma, WJ; Nobes, RH; Yates, BF "A Theoretical Approach to Gas-Phase Ion Chemistry," *Pure Appl. Chem.* **1984**, *56*, 1831.

60. Scalf, M; Westphall, MS; Smith, LM "Charge Reduction Electrospray Mass Spectrometry," *11th Sanibel Conference on Mass Spectrometry/Mass Spectrometry in Clinical Diagnosis of Disease*, January 23–26, 1999.

61. Sokolow, S; Karnofsky, J; Gustafson, P *Finnigan Application Report No. 2*; Finnigan: San Jose, CA, March 1978.

62. Stein, S, et al. "NIST/EPA/NIH Mass Spectral Database" and "NIST 98 Mass Spectral Search Program, v 1.6," Mass Spectrometry Data Center, National Institute of Standards and Technology, U.S. Department of Commerce, Gaithersburg, MD, source of mass spectra.

63. Stephenson, JL, Jr.; McLuckey, SA "Charge Reduction of Oligonucleotide Anions Via Gas-phase Electron Transfer to Xenon Cations," *Rapid Commun. Mass Spectrom.* **1997**, *11*(8), 875–880.

64. Stevenson, DP "Title Unknown," *Disc. Faraday Soc.* **1951**, *10*, 35.

65. Sundqvist, B; Macfarlane, RD "252-Cf Plasma Desorption Mass Spectrometry," *Mass Spectrom. Rev.* **1985**, *4*, 421–460.

66. Tal'roze, VL; Lyubimova, AK "Secondary Processes in a Mass Spectrometer Ion Source," *Doki Akad. Nauk SSSR* **1952**, *86*, 909.

67. Todd, JFJ International Union of Pure and Applied Chemistry, Physical Chemistry Division, Commission on Molecular Structure and Spectroscopy [*sic*], Subcommittee on Mass Spectroscopy [*sic*], "Recommendations For Nomenclature And Symbolism For Mass Spectroscopy" [*sic*] (including an appendix of terms used in vacuum technology) (*Current IUPAC Recommendations*, 1991); Todd, JFJ *Int. J. of Mass Spectrom. Ion Proc.* **1995**, *142*(3), 211–240.

68. U.S. Government *United States Government Printing Office Manual-Of-Style*; United States Government Printing Office: Washington, DC, 1986; p 134.

69. Watson, JT *Introduction to Mass Spectrometry*, 3rd ed.; Lippincott-Raven: Philadelphia-New York, 1997; p 118.

70. Whitehouse, CM; Fenn, JB; Yamashita, M "An Electrospray Ion Source for Mass Spectrometry of Fragile Organic Species," *Proceedings of the 32nd ASMS Conference on Mass Spectrometry and Allied Topics*, San Antonio, Texas, May 27–June 1, 1984; pp 188, 189.

71. Willoughby, RC; Browner, RF "Monodisperse Aerosol Generation Interface for LC/MS," *Anal. Chem.* **1984**, *56*, 2626.

72. Wilm, M; Mann, M "Analytical Properties of the Nanoelectrospray Ion Source," *Anal. Chem.* **1996**, *68*, 1.

73. Yergey, J; Heller, D; Hansen, G; Cotter, RJ; Fenselau, C "Isotope Dilution in MS of Large Molecules," *Anal. Chem.* **1983**, *55*, 353–356.

BIBLIOGRAPHY

Preparation of any such compilation is thwarted with the problem that by the time it reaches the intended audience, it will be out of date. As is the case with this effort, the generation of this collection of book titles was inspired to some degree by two such collections that are found in books authored by Roboz in 1968 (Reference 77) and Kiser in 1965 (Reference 81). I found both of these collections to have been invaluable in my study of mass spectrometry. More recent collections of book titles in mass spectrometry appeared in McLafferty and Tureček (Interpretation 6) and de Hoffman (Introductory 8). Both of these collections, like the Roboz and Kiser collections are dated, even though both were prepared in this decade.

The old adage, "You can't judge a book by its cover!" is truer today than ever before. Although there has been a number of outstanding mass spectrometry books that address the new technologies developed at the end of the twentieth century, the literature of mass spectrometry has been cursed with a plethora of recently published books that suffer from bad technical content, little or no competent copyediting, and/or amateurish and sloppy production. These problems are due to consolidations in the publishing industry resulting in fewer publishers, a perceived need for new books to replace older ones and to provide information on a subject that has exploded in the last decade with new ionization techniques, and the lack of desire and ability of authors to research their subject and avoid the self-gratification of generating an avalanche of neologisms and technical errors. These types of problems were practically nonexistent in books on mass spectrometry before 1990.

There is a well-written and organized general text on mass spectrometry (Introductory 4) that suffers from an incompetent production effort. In this particular book, there is a lack of consistency in the fonts used for the presentation of reaction schemes, figures are improperly imported from electronic submissions, and promised graphics on the inside cover are replaced with an easy-to-lose insert. This book is the singularly best modern text on mass spectrometry partially because its thousands of journal references all have titles. It is a shame that the publisher did not make the effort to do a better job of the presentation. Poor production efforts by established publishers is one of the reasons why you see more self-publishing (Introductory 2 and Technique 10).

Another book, produced by a major publisher, consistently uses the word "spectra" (the plural form) for the word "spectrum" (the singular form) (Technique 6). As a corollary to you can't judge a book by its cover, "you can't necessarily judge a book by its title." A book was published with the title *Understanding Mass Spectra: A Basic Approach* (Technique 3) when, in reality, it should have been entitled *How I Came to Understand the Mass Spectra of Illicit Drugs: an Autobiographical Presentation*. There are also problems with poorly translated foreign-language books such as one originally written in French and translated into English by a person whose native language was Chinese (Introductory 8); reviewed in *J. Am. Soc. Mass Spectrom.* **1997**, *8*, 1193, 1194. Another problem with translated books is the delay in information. This dated information is especially a problem with the rapid change that is taking place in LC/MS and mass spectrometry and biotechnology.

I have reviewed several books with copyright dates after 1997 for the *Journal of the American Society for Mass Spectrometry*.   One of the books is in the **Technique-Oriented Books** section (Technique 6: *J. Am. Soc. Mass Spectrom.* **1999**, *10*, 364–367) and one is in the **Interpretation Books** section (Interpretation 3: *J. Am. Soc. Mass Spectrom.* **1998**, *9*, 852–854).   Both of these books were found to be of little or no value, and it was felt that they would be more harmful to the reader than helpful. These two books should be avoided.

Most of the books on mass spectrometry come from the chemists who use the technique. Chemistry, unlike biochemistry and the biological sciences, has been slow to recognize the importance of the titles of journal articles in cited references.   Most of the mass spectrometry books published in the last 10 years (and nearly all those previously published) do not include titles with cited journal articles.   That is why a special effort has been made to mark books that do use titles in cited journal articles in this bibliography. These books are noted with an asterisk (*).

This bibliography is presented in 11 segments.   Book-related segments include: **Introductory**,   **Reference**,   **Technique-Oriented**,   **Interpretation**,   **Historical Significance**, **Collections of Mass Spectra in Hardcopy**, and **Integrated Spectral Interpretation**. Non-book-related segments include: **Mass Spectrometry, GC/MS, and LC/MS Journals**; **Personal Computer MS Abstract Sources**; **Software**; and **Monographs**.   These segments are included because of their importance in finding information on mass spectrometry.

There is no inclusion of specific articles from journals. Through the 1960s, such listings of journal articles were published in various forms—often by mass spectrometry instrumentation manufacturers.   However, since the development of comprehensive abstract systems and their electronic availability, these printed listings no longer have much value.   The **Personal Computer MS Abstract Sources** section is a guide to information on current journal articles.

There is no inclusion of Internet search engines, such as Medline and Chemical Abstracts Service. Some of these search engines are found on the **Important World Wide Web URLs** page on the back cover of this book.

There has been some duplication of book titles.   For example, the Watson book (Introductory 4 and Interpretation 4) covers two areas of mass spectrometry. This book is excellent for the interpretation of electron ionization spectra as well as an introductory book. All books are listed in chronological order by section.

Introductory Books

This section contains introductory books on mass spectrometry and gas and liquid chromatography. If you are going to own only one mass spectrometry book, it should be *Introduction to Mass Spectrometry*, 3rd edition (Introductory 4), which is a comprehensive book that touches on all the current technology. This book provides an excellent understanding of electron ionization (EI) fragmentation mechanisms (which is essential to the understanding of all fragmentation in mass spectrometry); and, of great importance, all of the journal articles referenced have titles. Another important book in this section is *Pushing Electrons: A Guide for Students of Organic Chemistry* (Introductory 3). This student workbook instills a good understanding of moving electrons within organic ions and molecules, which is also essential to the understanding of mass spectral fragmentation.

MS Fundamentals (Introductory 9) is not a book. It is a multimedia training tool consisting of a computer-based training program, video, and book. This entry is also listed in the **Software** section. To those who work with the transmission-quadrupole mass spectrometer, this multimedia presentation is indispensable. In order to get the most from any analytical instrument, you must have a thorough understanding of the technology. This multimedia presentation will result in that understanding, even in those who have little technical background.

There are three *Analytical Chemistry by Open Learning* books (Introductory 1, 10, and 16) listed in this section. These books are programmed-learning text and are very good for self-study. Two of the entries in this section are video courses produced by the American Chemical Society (Introductory 15 and 19). Neither is recommended because the mass spectrometry course is dated, and the gas chromatography course has too little emphasis on capillary columns.

1. Barker, J *Mass Spectrometry: Analytical Chemistry by Open Learning*, 2nd ed.; Ando, DJ, Ed.; Wiley: Chichester, U.K., 1999. (Davis, R; Frearson, MJ, 1st ed., 1987)

2. Cunico, RL; Gooding, KM; Wehr, T *Basic HPLC and CE of Biomolecules*; Bay Bioanalytical Laboratory: Richmond, CA, 1998.

3. Weeks, DP *Pushing Electrons: A Guide for Students of Organic Chemistry,* 3rd ed.; Saunders College: Fort Worth, TX, 1998.

4. *Watson, JT *Introduction to Mass Spectrometry*, 3rd ed.; Lippincott-Raven: Philadelphia-New York, 1997.

5. McNair, HM; Miller, JM *Basic Gas Chromatography*; Wiley: New York, 1997.

6. Siuzdak, G *Mass Spectrometry for Biotechnology*; Academic: San Diego, CA, 1996.

7. Johnstone, RAW; Rose, ME *Mass Spectrometry for Chemists and Biochemists*, 2nd ed.; Cambridge University Press: Cambridge, U.K., 1996.

8. de Hoffmann, E; Charette, J; Stroobant, V *Mass Spectrometry: Principles and Applications*; © Masson éditeur, Paris, 1996; Wiley: New York, 1996. (original French language edition, *Spectrométrie de masse*, © Masson éditeur, Paris, 1994)

9. *MS Fundamentals: Multimedia Training*; SAVANT: Fullerton, CA, 1995.

10. Fowlis, IA *Gas Chromatography: Analytical Chemistry by Open Learning*, 2nd ed.; Wiley: Chichester, U.K., 1995.

11. Russell, DH, Ed. *Experimental Mass Spectrometry*; Plenum: New York, 1994.

12. Hinshaw, JV; Ettre, LS *Introduction to Open-Tubular Column Gas Chromatography*; Advanstar: Cleveland, OH, 1993.

13. Ettre, LS; Hinshaw, JV *Basic Relationships of Gas Chromatography*; Advanstar: Cleveland, OH, 1993.

14. Chapman, JR *Practical Organic Mass Spectrometry: A Guide for Chemical and Biochemical Analysis*; Wiley: New York, 1993.

15. Watson, JT *Introduction to Mass Spectrometry*, ACS Video Courses; American Chemical Society: Washington, DC, 1993; 2 tapes, 130 pp.

16. Lindsay, S *High Performance Liquid Chromatography: Analytical Chemistry by Open Learning*, 2nd ed.; Wiley: Chichester, U.K., 1992. (1st ed., 1987)

17. *Desiderio, DM, Ed. *Mass Spectrometry: Clinical and Biomedical Applications*, Vols. 1 and 2; Plenum: New York, 1992.

18. Karasek, FW; Clement, RE *Basic Gas Chromatography-Mass Spectrometry*; Elsevier: New York, 1988.

19. McNair, HM *Basic Gas Chromatography*, ACS Video Courses; American Chemical Society: Washington, DC, 1988.

20. Beynon, JH; Brenton, AG *Introduction to Mass Spectrometry*; University of Wales Press: Swansea, U.K., 1982.

21. Johnson, E; Stevenson, R *Basic Liquid Chromatography*; Varian: Palo Alto, CA, 1978.

22. Snyder, LR; Kirkland, JJ *Introduction to Modern Liquid Chromatography*; Wiley: New York, 1978.

23. McNair, HM; Bonelli, EJ *Basic Gas Chromatography*, 5th ed.; Varian: Palo Alto, CA, 1969.

Reference Books

The citations in this section are meant to be general references on mass spectrometry. In addition, there is a reference to LC method development. One of the most important entries in this section is the "Mass Spectrometry" article in the biennial Fundamental Reviews issue of *Analytical Chemistry*. This review began in 1949 with the article authored by John A. Hipple and Martin Shepherd (National Bureau of Standards, currently the National Institute of Standards and Technology) with 165 citations. The latest of these reviews (Reference 5) had 1,551 citations divided into nine categories: Overview (5), Scope (173), Innovative Techniques and Instrumentation (364), Isotope-Ratio Mass Spectrometry (89), High-Power Laser in Mass Spectrometry (51), Dissociation by Low-Intensity Infrared Radiation (18), Polymers (61), Peptides and Proteins (624), and Oligonucleotides and Nucleic Acids (166). The "Mass Spectrometry" articles in the Fundamental Reviews issues of *Analytical Chemistry* have had Alma L. Burlingame as their primary author since 1972 with 14 consecutive articles.

An important aspect of using a mass spectrometer is "becoming one" with the instrument. In order to accomplish this unison, you must have an in-depth working understanding of the instrument. The multimedia training program distributed by SAVANT, *MS Fundamentals* (Introductory 9 and Software 23), is an excellent aid to gaining this understanding of the transmission-quadrupole mass spectrometer. In addition, the Dawson book (Reference 63) is a seminal reference for the transmission quadrupole and the quadrupole ion trap. The March/Hughes book (Reference 35) is the seminal reference for the quadrupole ion trap, and the Cotter book (Reference 12) is a likewise reference for the time-of-flight mass spectrometer. Care must be taken with respect to the Cotter title because he has edited a book (Reference 20) that does not provide the detailed understanding of the TOF-MS. The same is true for the March/Hughes book because of a similar title edited by March and Todd (Reference 17). There is no good reference for the current technology of magnetic-sector mass spectrometers. The best material on understanding the fundamentals of a magnetic-sector and/or double-focusing instrument is the Roboz book (Reference 77). The best understanding of the workings of FTMS is found in an article, "Fourier Transform Ion Cyclotron Resonance Mass Spectrometry: A Primer," authored by Allen Marshall et al. in *Mass Spectrom. Rev.* **1998**, *17*(1), 1–36.

Working with mass spectrometers often requires the derivatization of analytes to obtain the best results. This section has three good books that aid in this task (Reference 25, 53, and 78). In addition, there are two other books that do not primarily pertain to mass spectrometry techniques: *Basic Vacuum Practice* from Varian (Reference 36) and *The Mass Spec Handbook of Service* by Manura (Reference 24).

It should be noted that several books of the 1960s and early 1970s have been reprinted in the past few years (Reference 63, 82, and 84) by the American Institute of Physics and the American Society for Mass Spectrometry (ASMS). Republication is not a new practice. Several books were reprinted by a publisher other than the original publisher several years after the first printing. ASMS plans to reprint several other books of both historical and technological significance over the next few years.

1. *Hill, SJ, Ed. *ICP Spectrometry and Its Applications*; Sheffield Academic: Sheffield, U.K., 1999.

2. Shepard, KW, Ed. *Heavy Ion Accelerator Technology*: Eighth International Conference (AIP Conference Proceedings), Vol. 473; American Institute of Physics: New York, 1999.

3. *Sparkman, OD (p 2604); Wells, GJ and Huston, CK (p 2662); Adams, F, et al. p 2650). In *Encyclopedia of Environmental Analysis and Remediation*, Vol. 4; Meyers, RA, Ed.; Wiley: New York, 1998.

4. Karjalainen, EJ; Hesso, AE; Jalonen, JE; Karjalainen, UP, Eds. *Advances in Mass Spectrometry*, Proceedings of the 14th International Conference, 1997, Tampere, Finland; Elsevier: Amsterdam, 1998. (also available as a CD ROM)

5. Burlingame, AL; Boyd, RK; Gaskell, SJ "Mass Spectrometry" in the Fundamental Reviews issue of *Anal. Chem.* **1998**, *70*(16).

6. Tuniz, C; Tuniz, JR; Fink Bird, D, Eds. *Accelerator Mass Spectrometry: Ultrasensitive Analysis for Global Science*; CRC: Boca Raton, FL, 1998.

7. Grove, HE *From Hiroshima to the Iceman: The Development and Applications of Accelerator Mass Spectrometry*; American Institute of Physics: New York, 1998.

8. Montaser, A, Ed. *Inductively Coupled Plasma Mass Spectrometry*; VCH: Berlin, 1998.

9. Platzner, LT; Habfast, K; Walder, A; Goetz, A *Modern Isotope Ratio Mass Spectrometry*; Wiley: New York, 1997.

10. Holland, G; Tanner, SD, Eds. *Plasma Source Mass Spectrometry: Development and Applications*: International Conference on Plasma Source Mass Spectrometry; American Chemical Society: Washington, DC, 1997.

11. Wilkins, CL, Ed. Mass Spectrometry. In *Handbook of Instrumental Techniques for Analytical Chemistry*; Settle, FA, Ed.; Prentice Hall: Upper Saddle River, NJ, 1997; Section V.

12. Cotter, RJ *Time-of-Flight Mass Spectrometry: Instrumentation and Applications in Biological Research*; American Chemical Society: Washington, DC, 1997.

13. Snyder, LR; Kirkland, JJ; Glajch, JL *Practical HPLC Method Development*, 2nd ed.; Wiley: New York, 1997.

14. Ashcroft, AE *Ionization Methods in Organic Mass Spectrometry*, RSC Analytical Spectroscopy Monographs; Royal Society of Chemistry: Cambridge, U.K., 1997.

15. Baer, T; Ng, C-Y; Powis, I, Eds. *Large Ions: Their Vaporization, Detection and Structural Analysis*; Wiley: New York, 1996.

16. Townshead, A, et al., Eds. *Encyclopedia of Analytical Science*; Academic: San Diego, CA, 1995.

17. March, RE; Todd, JFJ, Eds. *Practical Aspects of Ion Trap Mass Spectrometry*, Vols. I, II, and III; CRC: Boca Raton, FL, 1995.

18. Ghosh, PK *International Series of Monographs on Physics, 90: Ion Traps*; Oxford Science: New York, 1995.

19. Corndies, I; Horváth, Gy; Vékey, K, Eds. *Advances in Mass Spectrometry*, Proceedings of the 13th International Conference, 1995, Budapest, Hungary; Elsevier: Amsterdam, 1995.

20. Cotter, RJ, Ed. *Time-of-Flight Mass Spectrometry*, ACS Symposium Series 549; American Chemical Society: Washington, DC, 1994.

21. Schlag, EW, Ed. *Time-of-Flight Mass Spectrometry and its Applications*; Elsevier: Amsterdam, 1994.

22. Matsuo, T; Caprioli, RM; Gross, ML; Seyama, Y, Eds. *Biological Mass Spectrometry: Present and Future*; Wiley: New York, 1994.

23. Benninghoven, A; Shimizu, R; Werner, HW; Nihei, Y *Secondary Ion Mass Spectrometry: SIMS IX*; Wiley: New York, 1994.

24. Manura, JJ; Baker, CW, Eds. *The Mass Spec Handbook of Service*, Vol. 2; Scientific Instrument Services: Ringoes, NJ, 1993.

25. Blau, K; Halket, J, Eds. *Handbook of Derivatives for Chromatography*; Wiley: New York, 1993.

26. Vertes, A; Gijbels, R; Adams, F *Laser Ionization Mass Analysis*; Wiley: New York, 1993.

27. Kistemaker, PG; Nibbering, NM, Eds. *Advances in Mass Spectrometry*, Proceedings of the 12th International Conference, Vol. 12, Amsterdam, 26–30 August 1991; Elsevier: Amsterdam, 1992.

28. Standing, KG; Ens, W, Eds. *Methods and Mechanisms for Producing Ions from Large Molecules*; Plenum: New York, 1991.

29. Jarvis, KE; Gray, AL; Houk, RS *Handbook of Inductively Coupled Plasma Mass Spectrometry*; Chapman Hall: London, 1991.

30. Marshall, AG; Verdun, FR *Fourier Transforms in NMR, Optical and Mass Spectrometry: A User's Handbook*; Elsevier: Amsterdam, 1990.

31. Jarvis, KE; Gray, AL; Williams, JG, Eds. *Plasma Source Mass Spectrometry*: The Proceedings of the Third Surrey Conference on Plasma Source Mass Spectrometry; Royal Society of Chemistry: London, 1990.

32. Constantin, E; Schnell, A (Chalmers, MH, Translator) *Mass Spectrometry*; Ellis Horwood: Chichester, U.K., 1990. (original French language edition, *Spectrométrie de masse*; Tec & Doc, France, © the copyright holders)

33. Lubman, DM, Ed. *Lasers in Mass Spectrometry*; Oxford University Press: Oxford, U.K., 1990.

34. Meuzelaar, HLC; Isenhour, TL, Eds. *Computer-Enhanced Analytical Spectroscopy*, Vol. 2; Plenum: New York, 1990.

35. *March, RE; Hughes, RJ *Quadrupole Storage Mass Spectrometry*; Wiley: New York, 1989.

36. Varian Vacuum Products Division *Basic Vacuum Practice*, 2nd ed.; Varian: Palo Alto, CA, 1989.

37. Benninghoven, A, Ed. *Ion Formation from Organic Solids: Mass Spectrometry of Involatile Materials*; Wiley: New York, 1989.

38. Heinrich, N; Schwarz, H. In *Ion and Cluster Ion Spectroscopy and Structure*; Maier, JP, Ed.; Elsevier: Amsterdam, 1989.

39. Wilson, RG; Stevie, FA; Magee, CW *Secondary Ion Mass Spectrometry: A Practical Handbook for Depth Profiling and Bulk Impurity Analysis*; Wiley: New York, 1989.

40. Prokai, L *Field Desorption Mass Spectrometry*; Marcel Dekker: New York, 1989.

41. Middleditch, BS *Analytical Artifacts: GC, MS, HPLC, TLC, and PC*, Journal of Chromatography Library, Vol. 44; Elsevier: Amsterdam, 1989.

42. Buchanan, MV, Eds. *Fourier Transform Mass Spectrometry*; American Chemical Society: Washington, DC, 1987.

43. Gray, NAB *Computer-Assisted Structure Elucidation*; Wiley: New York, 1986.

44. Futrell, JH, Ed. *Gaseous Ion Chemistry and Mass Spectrometry*; Wiley: New York, 1986.

45. Duckworth, HE; Barber, RC; Venkalasubramanian, VS *Mass Spectroscopy*, 2nd ed.; Cambridge University Press: Cambridge, U.K., 1986.

46. White, FA; Wood, GM *Mass Spectrometry: Applications in Science and Engineering*; Wiley: New York, 1986.

47. Mark, TD; Dunn, GH *Electron Impact Ionization*; Springer-Verlag: Berlin, Germany, 1985.

48. Beynon, JH; McGlashan, ML, Eds. *Current Topics in Mass Spectrometry and Chemical Kinetics*; Heyden: London, 1982.

49. Benninghoven, A *Secondary Ion Mass Spectrometry: Sims III*; Springer-Verlag: Berlin, 1982.

50. de Mayo, P, Ed. *Rearrangements in Ground and Excited States*, Vols. 1–3; Academic: New York, 1980.

51. Merritt, C, Jr.; McEwen, CN, Eds. *Practical Spectroscopy Series*, Vol. 3 *Mass Spectrometry: Part B*; Marcel Dekker: New York, 1980.

52. Merritt, C, Jr.; McEwen, CN, Eds. *Practical Spectroscopy Series*, Vol. 3 *Mass Spectrometry: Part A*; Marcel Dekker: New York, 1979.

53. Knapp, DR *Handbook of Analytical Derivatization Reactions*; Wiley-Interscience: New York, 1979.

54. Bowers, MT, Ed. *Gas Phase Ion Chemistry*, Vols. 1, 2, 1979 and Vol. 3, 1984; Academic: New York.

55. Franklin, JL, Ed. *Benchmark Papers in Physical Chemistry and Chemical Physics*, Vol. 3 *Ion-Molecule Reactions, Part I: The Nature of Collisions and Reactions of Ions with Molecules and Ion-Molecule Reactions*; *Part II: Elevated Pressures and Long Reaction Times*; Dowden, Hutchingson & Ross: Stroudsburg, PA, 1979.

56. Ausloos, PJ, Ed. *Kinetics of Ion-Molecule Reactions*, Nato Advanced Studies Institute Series; Plenum, New York, 1979.

57. Gross, ML, Ed. *High Performance Mass Spectrometry: Chemical Applications*, ACS Symposium Series 70; American Chemical Society: Washington, DC, 1978.

58. Levsen, K *Fundamental Aspects of Organic Mass Spectrometry*; Verlag Chemie: Weinheim, Germany, 1978.   Note: This book is Volume 4 of a series entitled *Progress in Mass Spectrometry Fortschritte der Massenspektrometrie* edited by Herausegeben von Herbert Budzikiewicz. Vol. 1: Hesse, M *Indolakaloide*, Teil 1 (Text), Teil 2 (Spektren); Vol. 2: Drewes, SE *Chroman and Related Compounds*; Vol. 3: Hesse, M; Bernhard, HO *Alkaloide (außer Indol-, Triterpen- und Steroidalkaloide)*. The publication dates of these three previous volumes is not known, nor is it known if there are subsequent volumes.

59. Millard, BJ *Quantitative Mass Spectrometry*; Heyden: London, 1978.

60. Cooks, RG, Ed. *Collision Spectroscopy*; Plenum: New York, 1978.

61. Majer, JR *The Mass Spectrometer*; Taylor and Francis: Bristol, PA, 1977.

62. Gudzinowicz, BJ; Gudzinowicz, MJ; Martin, HF *Fundamentals of Integrated GC-MS* (in three parts), *Part III: The Integrated GC-MS Analytical System*; Marcel Dekker: New York, 1977.

63. Dawson, PH, Ed. *Quadrupole Mass Spectrometry and Its Applications*; Elsevier: Amsterdam, 1976. (reprinted by American Institute of Physics: Woodbury, NY, 1995)

64. Gudzinowicz, BJ; Gudzinowicz, MJ; Martin, HF *Fundamentals of Integrated GC-MS* (in three parts), *Part II: Mass Spectrometry*; Marcel Dekker: New York, 1976.

65. Gudzinowicz, BJ; Gudzinowicz, MJ; Martin, HF *Fundamentals of Integrated GC-MS* (in three parts), *Part I: Gas Chromatography*; Marcel Dekker: New York, 1976.

66. Lehman, TA; Bursey, MM *Ion Cyclotron Resonance Spectrometry*; Wiley: New York, 1976.

67. Lias, SG; Ausloos, PJ *Ion-Molecule Reactions: Their Role in Radiation Chemistry*; American Chemical Society: Washington, DC, 1975.

68. Waller, GR, Ed. *Biochemical Applications of Mass Spectrometry*; Wiley-Interscience: New York, 1972. (Waller, GR; Dermer, OC, 1st supplement, 1980)

69. Williams, DH; Howe, I *Principles of Organic Mass Spectrometry*, 2nd ed.; McGraw-Hill: London, 1972. (1st ed., 1964)

70. Maccoll, A, Ed. *Mass Spectrometry*; MTP Int. Rev. Sci. Butterworths: London, 1972.

71. Franklin, JL, Ed. *Ion-Molecule Reactions*, Vol. 1 and Vol. 2; Plenum: New York, 1972.

72. Milne, GWA, Ed. *Mass Spectrometry: Techniques and Applications*; Wiley-Interscience: New York, 1971.

73. Williams, DH, Ed. *Mass Spectrometry*, Vol. 1, 1971 and Vol. 2, 1973; Chemical Society: London.

74. Field, FH; Franklin, JL *Electron Impact Phenomena and the Properties of Gaseous Ions*, revised edition; Academic: New York, 1970. (1st ed., 1957)

75. McDaniel, EW; Čhermák, V; Dalgarno, A; Ferguson, EE; Friedman, L *Ion-Molecule Reactions*; Wiley-Interscience: New York, 1970.

76. Knewstubb, PF *Mass Spectrometry and Ion-molecule Reactions*; Cambridge University Press: Cambridge, U.K., 1969.

77. Roboz, J *Introduction to Mass Spectrometry Instrumentation and Techniques*; Wiley: New York, 1968.

78. Pierce, AE *Silylation of Organic Compounds*; Pierce Chemical: Rockford, IL, 1968.

79. White, FA *Mass Spectrometry in Science and Technology*; Wiley: New York, 1968.

80. Ausloos, PJ, Ed. *Ion-Molecule Reactions in the Gas Phase*, ACS Advances in Chemistry Series 58; American Chemical Society: Washington, DC, 1966.

81. Kiser, RW *Introduction to Mass Spectrometry and Its Application*; Prentice-Hall: Englewood Cliffs, NJ, 1965.

82. Biemann, K *Mass Spectrometry: Organic Chemical Applications*; McGraw-Hill: New York, 1962. (reprinted by ASMS, 1998)

83. Reed, RI *Ion Production by Electron Impact*; Academic: London, 1962.

84. Beynon, JH *Mass Spectrometry and Its Applications to Organic Chemistry*; Elsevier: Amsterdam, 1960. (reprinted by ASMS, 1999)

Technique-Oriented Books

The books listed in this section pertain to specific techniques of mass spectrometry and hyphenated chromatography/mass spectrometry techniques. In some cases, the books are specific to certain types of analytes. Books of this type began to appear at the end of the 1960s (only one listing before 1970, Technique 91). Slightly more than half of these books (47 out of 91) have copyright dates in the last decade. Unfortunately, the last three or four years have seen a proliferation of very poorly written books with little or no copyediting and poor quality production.

Many of the technique-oriented books are made obsolete by changing technology within a few years or months of their publication. A good example is the 1990 Yergey book (Technique 39) on LC/MS. Because of the massive advancements in technology that have taken place in the last 10 years, this book had little relevance to the technique within 2 years of its publication. The book still has a great deal of value in that it provides good information on how to perform the chromatographic separations required in LC/MS, and it has an excellent set of journal-article references that contains titles. The problem of dated material can be especially significant with foreign-language books that are translated into English. Unfortunately, in some cases, publishers are not indicating that a book has been translated. It is only through careful research that the foreign-language roots of a book can be established, such as the case with Technique 2.

There are two books (Technique 10 and 21) that are of particular value to the LC/MS and GC/MS practitioner, respectively. These two books have a great deal of practical information on the running of different types of analyses and are good aids in the decision-making process about how to proceed with a particular sample. The Willoughby book (Technique 10) has very useful information in deciding whether to use a contract laboratory or perform the analysis in-house. The Willoughby book was reviewed in *JASMS* (*J. Am. Soc. Mass Spectrom.* **1999**, *10*, 78, 79) as was the Kitson book (*J. Am. Soc. Mass Spectrom.* **1997**, *9*, 294, 295). Care must be taken with respect to the unfortunate similarity between the title of the Kitson book (Technique 21) and the book title of Technique 6.

Books that are edited works rather than having a single author generally don't get my approval. These edited editions often end up looking like a "camel" (a horse designed by a committee). This lack of continuity in edited books is more true of books published in the last two decades than those published before that time. There is one notable exception in Technique 15 by Cole (reviewed *J. Am. Soc. Mass Spectrom.* **1997**, *8*, 1191, 1192). This book is an excellent reference for those working in electrospray. The second edition of the Niessen book (Technique 5) is also a good reference for electrospray as well as other LC/MS techniques. The single negative about both of these books is that they do not include titles with journal-article citations.

There are two important references on environmental GC/MS that should be reviewed by anyone working in this area (Technique 71 and 73). Although both of these books were written in the era of the packed column, the fundamentals of environmental analyses and the U.S. Environmental Protection Agency (EPA) tune-criteria are covered in detail.

Another reference of the packed-column era is the McFadden book (Technique 86). This book, along with the Karasek book (Introductory 18), is very useful to those starting in GC/MS.

If you are using chemical ionization (either atmospheric pressure chemical ionization or chemical ionization under the conditions normally encountered in GC/MS), you need the Harrison book (Technique 33). Unlike the Yinon book (Technique 27), this book is a second edition and is labeled as such.

In looking at new titles of technique-orientated books, care must be taken to know when the book is nothing more than a collection of a series of articles from a journal or a bound issue of a journal. The value of such books is often less than their extremely high selling price.

1. *Willoughby, R; Sheehan, E *A Global View of MS/MS*; Global View: Pittsburgh, PA, 2000.

2. Gerhards, P; Bons, U; Sawazki, J; Szigan, J; Wertmann, A *GC/MS in Clinical Chemistry*; Wiley: Chichester, U.K., 1999.

3. Smith, RM (Busch, KL, Tech. Ed.) *Understanding Mass Spectra: A Basic Approach*; Wiley: New York, 1999.

4. Montaudo, G; Lattimer, RP *Mass Spectrometry of Polymers*; CRC: Boca Raton, FL, 1999.

5. Niessen, WMA *Liquid Chromatography-Mass Spectrometry*, 2nd ed., Chromatographic Science Series, Vol. 79; Marcel Dekker: New York, 1998.

6. McMaster, M; McMaster, C *GC-MS: A Practical User's Guide*; Wiley: New York, 1998.

7. Niessen, WMA; Voyksner, RD, Eds. *Current Practice in Liquid Chromatography-Mass Spectrometry*, reprinted from *Journal of Chromatography A*, Vol. 794; Elsevier: Amsterdam, 1998.

8. Ens, W; Standing, KG; Chernushevich, IV, Eds. *New Methods for the Study of Biomolecular Complexes*, A NATO ASI Series (Series C), Mathematical and Physical Science, Proceedings of the NATO Advanced Research Workshop on New Methods for the Study of Molecular Aggregates, The Lodge at Kananaskis Village, Alberta, Canada, 16–20 June 1996, Vol. 510; Kluwer Academic: Boston, MA, 1998.

9. *Larsen, BS; McEwen, CN, Eds. *Mass Spectrometry of Biological Materials*; Marcel Dekker: New York, 1998.

10. *Willoughby, R; Sheehan, E; Mitrovich, S *A Global View of LC/MS*; Global View: Pittsburgh, PA, 1998.

11. *Thurman, EM; Mills, MS *Solid Phase Extraction: Principles and Practice*; Wiley: New York, 1998.

12. Meyers, RA, Ed. *Encyclopedia of Environmental Analysis and Remediation*; Wiley: New York, 1998.

13. Briggs, D; Ward, IM; Suresh, S; Clarke, DR, Eds. *Surface Analysis of Polymers by XPS and Static SIMS*; Cambridge University Press: Cambridge, U.K., 1998.

14. Caprioli, RM; Malorni, A; Sindona, G, Eds. *Selected Topics in Mass Spectrometry in the Biomolecular Sciences: A Tutorial*, A NATO ASI Series (Series C), Mathematical and Physical Science, Vol. 504, Altavilla-Milicia (PA), Italy, 7–18 July 1996; Kluwer Academic: Boston, MA, 1997.

15. Cole, RB, Ed. *Electrospray Ionization Mass Spectrometry: Fundamentals, Instrumentation, and Applications*; Wiley: New York, 1997.

16. Newton, RP; Walton, TJ, Eds. *Proceedings of the Phytochemical Society of Europe, 40: Applications of Modern Mass Spectrometry in Plant Science Research*; Clarendon: Oxford, U.K., 1997.

17. Hancock, WS *New Methods in Peptide Mapping for the Characterization of Proteins*; CRC: Boca Raton, FL, 1996.

18. Barcelo, D, Ed. *Applications of LC-MS in Environmental Chemistry*, Journal of Chromatography Library, Vol. 59; Elsevier: Amsterdam, 1996.

19. Burlingame, AL; Carr, SA, Eds. *Mass Spectrometry in the Biological Sciences*; Humana: Totowa, NJ, 1996.

20. *Chapman, JR, Ed. *Protein and Peptide Analysis by Mass Spectrometry*; Humana: Totowa, NJ, 1996.

21. *Kitson, FG; Larsen, BS; McEwen, CN *Gas Chromatography and Mass Spectrometry: A Practical Guide*; Academic: San Diego, CA, 1996.

22. Caprioli, RM; Malorni, A; Sindona, G, Eds. *Mass Spectrometry in the Biomolecular Sciences: A Tutorial*, A NATO ASI Series (Series C), Mathematical and Physical Science, Vol. 475, Lacco Ameno, Ischia, Italy, June 23–July 5, 1993; Kluwer Academic: Boston, MA, 1996.

23. Karger, BL; Hancock, WS, Eds. *Methods in Enzymology*, Vols. 270 and 271 *High Resolution Separation and Analysis of Biological Macromolecules*, *Part A: Fundamentals* and *Part B: Applications*; Academic: San Diego, CA, 1996. (two separate books)

24. *Boutton, TW; Yamasaki, S-i, Eds. *Mass Spectrometry of Soils*; Marcel Dekker: New York, 1996.

25. *Walker, JM, Ed. *The Protein Protocols Handbook*; Humana: Totowa, NJ, 1996.

26. Snyder, AP, Ed. *Biochemical and Biotechnology Applications of Electrospray Ionization Mass Spectrometry*, ACS Symposium Series 619; American Chemical Society: Washington, DC, 1995.

27. Yinon, J, Ed. *Forensic Applications of Mass Spectrometry*; CRC: Boca Raton, FL, 1995.

28. Fenselau, C, Ed. *Mass Spectrometry for the Characterization of Microorganisms*, ACS Symposium Series 549; American Chemical Society: Washington, DC, 1994.

29. Ardrey, B, Ed. *Liquid Chromatography/Mass Spectrometry*; VCH: New York, 1993.

30. *Murphy, RC *Handbook of Lipid Research*, No. 7 *Mass Spectrometry of Lipids*; Plenum: New York, 1993.

31. Niessen, WMA; van der Greef, J *Liquid Chromatography-Mass Spectrometry*, Chromatographic Science Series, Vol. 58; Marcel Dekker: New York, 1992.

32. Gross, ML, Ed. *Mass Spectrometry in the Biological Sciences: A Tutorial*, A NATO ASI Series; Kluwer Academic: Boston, MA, 1992.

33. *Harrison, AG *Chemical Ionization Mass Spectrometry*; CRC: Boca Raton, FL, 1992.

34. St. Pyrek, J *Mass Spectrometry in the Chemistry of Natural Products. In *Recent Advances in Phytochemistry*, Vol. 25 *Modern Phytochemical Methods*; Fischer, NH, et al., Eds.; Plenum: New York, 1991; Chapter 6.

35. Czanderna, AW; Hercules, DM *Ion Spectroscopies for Surface Analysis*; Plenum: New York, 1991.

36. Hilf, ER, Ed. *Mass Spectrometry of Large Non-Volatile Molecules for Marine Organic Chemistry*; World Scientific: River Edge, NJ, 1990.

37. *Fox, A; Morgan, SL; Larsson, L; Odham, G, Eds. *Analytical Microbiology Methods: Chromatography and Mass Spectrometry*; Plenum: New York, 1990.

38. Burlingame, AL; McCloskey, JA, Eds. *Biological Mass Spectrometry*, Proceedings of the 2nd International Symposium of Mass Spectrometry in Health & Life Sciences, San Francisco, CA, August 27–31, 1989; Elsevier: Amsterdam, 1990.

39. *Yergey, AL; Edmonds, CG; Lewis, IAS; Vestal, ML *Liquid Chromatography/Mass Spectrometry: Techniques and Applications*; Plenum: New York, 1990.

40. Halket, JM; Rose, ME *Introduction to Bench-Top GC/MS*; HD Science: Stapleford, U.K., 1990.

41. Suelter, CH; Watson, JT, Eds. *Methods of Biochemical Analysis*, Vol. 34 *Biomedical Applications in Mass Spectrometry*; Wiley-Interscience: New York, 1990.

42. McCloskey, JA, Ed. *Methods in Enzymology*, Vol. 193 *Mass Spectrometry*; Academic: San Diego, CA, 1990.

43. McEwen, CN; Larsen, BS, Eds. *Practical Spectroscopy Series: Mass Spectrometry of Biological Materials*; Marcel Dekker: New York, 1990.

44. Brown, MA, Ed. *Liquid Chromatography/Mass Spectrometry: Applications in Agricultural, Pharmaceutical and Environmental Chemistry*, ACS Symposium Series 420; American Chemical Society: Washington, DC, 1990.

45. Caprioli, RM, Ed. *Continuous-Flow Fast Atom Bombardment Mass Spectrometry*; Wiley: New York, 1990.

46. Desiderio, DM, Ed. *Mass Spectrometry of Peptides*; CRC: Boca Raton, FL, 1990.

47. SCIEX, *The API Book*; SCIEX, Division of MDS Health Group: Mississauga, Ontario, Canada, 1990.

48. Dolan, JW; Snyder, LR *Troubleshooting LC Systems*; Humana: Totowa, NJ, 1989.

49. Ashe, TR; Wood, KV, Eds. *Novel Techniques in Fossil Fuel Mass Spectrometry*; ASTM: Washington, DC, 1989.

50. Farrar, JM; Saunders, WH *Techniques for the Study of Ion-Molecule Reactions*, Vol. 20 *Techniques of Chemistry*; Wiley: New York, 1988.

51. Biermann, CJ; McGinnis, GD, Eds. *Analysis of Carbohydrates by GLC and MS*; CRC: Boca Raton, FL, 1988.

52. Adams, F *Inorganic Mass Spectrometry*, Chemical Analysis: A Series of Monographs on Analytical Chemistry and Its Applications; Wiley: New York, 1988.

53. Lai, S-TF *Gas Chromatography/Mass Spectrometry Operation*; Realistic Systems: East Longmeadow, MA, 1988.

54. Busch, KL; Glish, GL; McLuckey, SA, Eds. *Mass Spectrometry/Mass Spectrometry: Techniques and Applications of Tandem Mass Spectrometry*; VCH: New York, 1988.

55. McNeal, CJ, Ed. *The Analysis of Peptides and Proteins by Mass Spectrometry*, Proceedings of the 4th Texas Symposium, College Station, TX, April 17–20, 1988; Wiley: New York, 1988.

56. Gilbert, J, Ed. *Applications of Mass Spectrometry in Food Science*; Elsevier: London, 1987.

57. Heinzle, E; Reuss, M, Eds. *Mass Spectrometry in Biotechnological Process Analysis and Control*; Plenum: New York, 1987.

58. Rosen, JD, Ed. *Applications of New Mass Spectrometry Techniques in Pesticide Chemistry*, Chemical Analysis: A Series of Monographs on Analytical Chemistry and Its Applications, Vol. 91; Winefordner, JD, Ed.; Kolthoff, IM, Editor Emeritus; Wiley-Interscience: New York, 1987.

59. *Linskens, HF; Jackson, JF, Eds. *Modern Methods of Plant Analysis: Gas Chromatography/Mass Spectrometry*, New Series, Vol. 3; Springer-Verlag: Berlin, 1986.

60. McNeal, CJ, Ed. *Mass Spectrometry in the Analysis of Large Molecules*; Wiley: Chichester, U.K., 1986.

61. Gaskell, SJ, Ed. *Mass Spectrometry in Biomedical Research*; Wiley: Chichester, U.K., 1986.

62. Aczel, T, Ed. *Mass Spectrometric Characterization of Shale Oils*; ASTM: Philadelphia, PA, 1986.

63. Lyon, PA, Ed. *Desorption Mass Spectrometry: Are SIMS and FAB the Same?*; American Chemical Society: Washington, DC, 1985.

64. Karasek, FW; Hutzinger, O; Safe, S, Eds. *Mass Spectrometry in Environmental Sciences*; Plenum: New York, 1985.

65. Facchetti, S, Ed. *Mass Spectrometry of Large Molecules*; Elsevier: Amsterdam, 1985.

66. Message, GM *Practical Aspects of Gas Chromatography/Mass Spectrometry*; Wiley: New York, 1984.

67. Odham, G; Larsson, L; Mardh, P-A, Eds. *Gas Chromatography/Mass Spectrometry: Applications in Microbiology*; Plenum: New York, 1984.

68. McLafferty, FW, Ed. *Tandem Mass Spectrometry*; Wiley-Interscience: New York, 1983.

69. *Goodman, SI; Markey, SP *Diagnosis of Organic Academias By Gas Chromatography-Mass Spectrometry*, Vol. 6 *Laboratory and Research Methods in Biology and Medicine*; Alan R. Liss: New York, 1981.

70. Meuzelaar, HLC; Haverkamp, J; Hileman, SD *Pyrolysis Mass Spectrometry of Biomaterials*; Elsevier: Amsterdam, 1980.

71. Budde, WL; Eichelberger, JW *Organics Analysis Using Gas Chromatography/Mass Spectrometry*; Ann Arbor Science: Ann Arbor, MI, 1979.

72. Middleditch, BS, Ed. *Practical Mass Spectrometry*; Plenum: New York, 1979.

73. Keith, LH, Ed. *Identification & Analysis of Organic Pollutants in Water*; Ann Arbor Science: Ann Arbor, MI, 1979.

74. Frigerio, AF, Ed. *Recent Developments in Mass Spectrometry in Biochemistry and Medicine*, Vol. 2; Proceedings of the 5th International Symposium on Mass Spectrometry in Biochemistry and Medicine, Rimini, Italy, June 1978; Plenum: New York, 1979.

75. *Payne, JP; Bushman, JA; Hill, DW, Eds. *Medical and Biological Applications of Mass Spectrometry*; Academic: London, 1979.

76. Land, DG; Nursten, HE *Progress in Flavour Research*; Applied Science: London, 1979.

77. Millard, BJ *Quantitative Mass Spectrometry*; Heyden: London, 1978.

78. Frigerio, AF; Ghisalberti, EL, Eds. *Mass Spectrometry in Drug Metabolism*, Proceedings of the International Symposium on Mass Spectrometry in Drug Metabolism, Mario Negri Institute for Pharmacological Research, Milan, Italy, June 1976; Plenum: New York, 1977.

79. Gudzinowicz, BJ, Ed. *Analysis of Drugs and Metabolites by Gas Chromatography: Mass Spectrometry*, Vol. 1 *Respiratory Gases, Ethyl Alcohol, and Related Toxicological Materials*, 1977; Vol. 2 *Hypnotics, Anticonvulsants, and Sedatives;* Vol. 3 *Antipsychotics, Antiemetics, and Antidepressant Drugs*; Vol. 4 *Central Nervous System Stimulants*; Vol. 5 *Analgesics, Local Anaesthetics, and Antibiotics*, 1978; Vol. 6 *Cardiovascular, Antihypertensive, Hypoglycemic, and Tiered-Related Agents*, 1979; Vol. 7 *Subtitle Unknown*, 1980; Marcel Dekker: New York.

80. Beckey, HD *Principles of Field Ionization and Field Desorption Mass Spectrometry*; Pergamon: New York, 1977.

81. Masada, Y *Analysis of Essential Oils by Gas Chromatography and Mass Spectrometry*; Halsted (Division of Wiley): New York, 1976. (© 1976, Hirokawa: Japan)

82. Zaretskii, ZV *Mass Spectrometry of Steroids*; Wiley: New York, 1976.

83. *Frigerio, AF; Ghisalberti, EL, Eds. *Mass Spectrometry in Biochemistry and Medicine*, Monographs of the Mario Negri Institute for Pharmacological Research, Milan, Italy; Raven: New York, 1974.

84. Haque, R; Biros, FJ, Eds. *Mass Spectrometry and NMR Spectroscopy in Pesticide Chemistry*; Plenum: New York, 1974.

85. Cooks, RG; Beynon, JH; Caprioli, RM; Lester, GR *Metastable Ions*; Elsevier: New York, 1973.

86. McFadden, W *Techniques of Combined Gas Chromatography/Mass Spectrometry: Applications in Organic Analysis*; Wiley-Interscience: New York, 1973.

87. Costa, E; Holmstedt, B, Eds. *Gas Chromatography-Mass Spectrometry in Neurobiology*; Raven: New York, 1973.

88. Ahearn, AJ, Ed. *Trace Analysis by Mass Spectrometry*; Academic: New York, 1972.

89. Melton, CE *Principles of Mass Spectrometry and Negative Ions*; Marcel Dekker: New York, 1970.

90. Burlingame, AL; Castagnoli, N, Eds. *Topics in Organic Mass Spectrometry*; Wiley-Interscience: New York, 1970.

91. Ettre, LS; McFadden, WH, Eds. *Ancillary Techniques of Gas Chromatography*; Wiley: New York, 1969.

Interpretation Books

The seminal book for the interpretation of EI mass spectra is the McLafferty book (Interpretation 6). This book is extremely valuable, but may be too advanced for a beginner trying to self-teach. The beginner should try to start with either the McLafferty 2nd edition book (Interpretation 8) or the book by Shrader (Interpretation 1), both of which have been out of print for some time but can be found on the used-book Web sites.

The books by Budzikiewicz et al. (Interpretation 19, 23, and 24) were the first books written using mechanisms in organic reactions to describe the fragmentation of energetic ions produced by electron ionization. The information in these books still has a great deal of relevance to the subject. All of the interpretation books listed in this section pertain primarily to odd-electron molecular ions formed by the EI process. The ions formed in LC/MS are predominately protonated molecules, which are even-electron ions. Even-electron ions fragment to produce other even-electron ions, which requires the breaking of more than one bond in the ion. Many of the neutral losses in even-electron ion fragmentation and odd-electron ion fragmentation are the same.

One important note on Interpretation 3: Just as book titles can be similar (Technique 6 and 21), people can have the same or similar names that can result in confusion. The author of Interpretation 3, Terrence A. Lee, a Department of Chemistry faculty member at Middle Tennessee State University in Murfreesboro, TN, should not be confused with Terry Lee, a noted researcher in mass spectrometry of biological substances at the Beckman Research Institute, City of Hope/Division of Immunology in Duarte, CA.

1. Shrader, SR *Introductory Mass Spectrometry*, 2nd ed.; Shrader Laboratories: Detroit, MI, 1999. (originally published by Allyn and Bacon: Boston, MA, 1971)

2. Barker, J *Mass Spectrometry: Analytical Chemistry by Open Learning*, 2nd ed.; Ando, DJ, Ed.; Wiley: Chichester, U.K., 1999. (Davis, R; Frearson, MJ, 1st ed., 1987)

3. Lee, TA *A Beginner's Guide to Mass Spectral Interpretation*; Wiley: Chichester, U.K., 1998.

4. *Watson, JT *Introduction to Mass Spectrometry*, 3rd ed.; Lippincott-Raven: Philadelphia-New York, 1997.

5. Splitter, JS; Tureček, F, Eds. *Applications of Mass Spectrometry to Organic Stereochemistry*; VCH: New York, 1994.

6. McLafferty, FW; Tureček, F *Interpretation of Mass Spectra*, 4th ed.; University Science: Mill Valley, CA, 1993.

7. Porter, QN *Mass Spectrometry of Heterocyclic Compounds*, 2nd ed.; Wiley-Interscience: New York, 1985.

8. McLafferty, FW; Venkataraghavan, R *Mass Spectral Correlations*, 2nd ed.; American Chemical Society: Washington, DC, 1982.

9. Sklarz, B, Ed. *Mass Spectrometry of Natural Products*, plenary lecturers presented at the International Mass Spectrometry Symposium on Natural Products, Rehovot, Israel, 28 August—2 September 1977; Pergamon: Oxford, U.K., 1978.

10. Levsen, K *Fundamental Aspects of Organic Mass Spectrometry*; Verlag Chemie: Weinheim, Germany, 1978.

11. DeJongh, DC *Interpretation of Mass Spectra*, ACS Audio Series; American Chemical Society: Washington, DC, 1975; 6 audio cassette tapes, 158 pp.

12. McLafferty, FW *Interpretation of Mass Spectra*, 2nd ed.; Benjamin: Reading, MA, 1973.

13. Hamming, MG; Foster, NG *Interpretation of Mass Spectra of Organic Compounds*; Academic: New York, 1972.

14. Hill, HC *Introduction to Mass Spectrometry*, 2nd ed.; Heyden: London, 1972. (1st ed., 1966)

15. Seibl, J *Massenspektrometrie*; Akademische Verlagsgesellschaft: Frankfurt, Germany, 1970.

16. Brymner, R; Penney, JR, Eds. *Mass Spectrometry*; Chemical: New York, 1969.

17. Beynon, JH; Saunders, RA; Williams, AE *The Mass Spectra of Organic Molecules*; Elsevier: Amsterdam, 1968.

18. Polyakova, AA; Khmel'nitskii, RA (Schmorak, J, Translator) *Introduction to Mass Spectrometry of Organic Compounds*; Israel Program For ScientificTranslations: Jerusalem, Israel, 1968. (original Russian language edition, *Vvedenie V Mass Spektrometriyu Organicheskikh Soedinenii*, Izdatel'stvo "Khimya," Moskva-Leningrad, 1966).

19. Budzikiewicz, H; Djerassi, C; Williams, DH *Mass Spectrometry of Organic Compounds*; Holden-Day: San Francisco, CA, 1967.

20. Reed, RI *Applications of Mass Spectrometry to Organic Chemistry*; Academic: New York, 1966.

21. Spiteller, G *Massenspektrometrische Strukturanalyse Organischer Verbindunqen*; Verlag Chemie: Weinheim, Germany, 1966.

22. Quayle, A; Reed, RI Interpretation of Mass Spectra. In *Interpretation of Organic Spectra*; Mathieson, DW, Ed.; Academic: New York, 1965.

23. Budzikiewicz, H; Djerassi, C; Williams, DH *Structure Elucidation of Natural Products by Mass Spectrometry*, Vol. I *Alkaloids*; Vol. II *Steroids, Terpenoids, Sugars, and Miscellaneous Natural Products*; Holden-Day: San Francisco, CA, 1964.

24. Budzikiewicz, H; Djerassi, C; Williams, DH *Interpretation of Mass Spectra of Organic Compounds*; Holden-Day: San Francisco, CA, 1964.

25. McLafferty, FW, Ed. *Mass Spectra of Organic Ions*; Academic: New York, 1963.

26. Beynon, JH; Williams, AE *Mass and Abundance Tables for Use in Mass Spectrometry*; Elsevier: New York, 1963.

27. McLafferty, FW Mass Spectrometry. In *Determination of Organic Structures by Physical Methods*, Vol. II; Nachod, FC; Phillips, WD, Eds.; Academic: New York, 1962.

Books of Historical Significance*

Based on an 1886 paper (*Berl. Ber.* **1886**, *39*, 691) by Eugene Goldstein (German physicist, 1850–1930) reporting the discovery of luminous rays emerging as straight lines from holes in a metal disc used as a cathode in a discharge tube (he called the rays Kanalstrahlen: canal rays) and the confirmation by Wilhelm Carl Werner Otto Fritz Franz Wien (German Nobel Laureate in Physics, 1911, 1864–1928) that Jean Baptiste Perrin's (French physicist, 1870–1942) 1895 postulation that the rays were associated with positive charge by studying their deflection in electric and magnetic fields (*Verh. Phys. Ges.* **1898**, *17*, 1898; *Ann. Physik.* **1898**, *65*, 440; *Ann. Phys. Leipzig* **1902**, *8*, 224), the field of mass spectrometry developed into a science between 1911 and 1925. This development was due to the results of the experiments conducted by the three founding fathers of mass spectrometry: Joseph John Thomson (English Nobel Laureate in Physics, 1906, 1856–1940); Francis Williams Aston (English Nobel Laureate in Chemistry, 1922, 1877–1945; Aston was an associate of Thomson in the Cavendish Laboratory in Manchester, England); and Arthur Jeffery Dempster (1886–1950, Canadian-American physics professor, University of Chicago).

In his 1968 book, Roboz (Reference 77) lists 20 selected papers for those wanting to learn the history of mass spectrometry through original references (Chapter 14, p 490). Of these papers, five were authored by Aston, four by Dempster, and two by Thomson. Another three were authored by William R. Smythe (U.S. scientist) and two by Kenneth Bainbridge (U.S. physicist, 1904–1906, Director of the Trinity test—the first test explosion of the atomic bomb), who also were early pioneers in mass spectrometry.

1. Davis, EA, Falconer, IJ *J. J. Thomson and the Discovery of the Electron*; Taylor & Francis: London, 1997.

2. Dahl, PF *Flash of the Cathode Rays: A History of JJ Thomson's Electron*; American Institute of Physics: Philadelphia, PA, 1997.

3. Ausloos, P, Ed. *Kinetics of Ion-Molecule Reactions*, Proceedings of the NATO Advanced Study Institute on Kinetics of Ion-Molecule Reactions held at Biarritz, France, September 4–15, 1978, Vol. 40; published in conjunction with NATO Scientific Affairs Division; Plenum: New York, 1979.

4. Nachod, FC; Zuckerman, JJ; Randall, EW, Eds. *Determination of Organic Structures by Physical Methods*, Vol. 6; Academic: New York, 1976.

5. Ausloos, P, Ed. *Interactions Between Ions and Molecules*, Proceedings of the NATO Advanced Study Institute on Kinetics of Ion-Molecule Reactions held at La Baule, France, 1974; published in conjunction with NATO Scientific Affairs Division; Plenum: New York, 1975.

6. Frigerio, A, Ed. *Mass Spectrometry in Biochemistry and Medicine*, Vols. 1, 2 *Advances in Mass Spectrometry in Biochemistry and Medicine*, 1974, 1975; *Mass Spectrometry in Drug Metabolism*, 1976; Vols. 1, 2, 6, 7 and Vol. 8 *Recent Developments in Mass Spectrometry in Biochemistry and Medicine*, 1977–1980 and 1982; a series of books published by Plenum: New York and Elsevier: Amsterdam as the proceedings of a meeting organized by the Mario Negri Institute for Pharmacological Research in Milan, Italy.

---

* Historical references are in the Bibliography section of the Kiser book (Reference 81) and in the Information and Data chapter of the Roboz book (Reference 77).

7. Biennial Specialist Periodical Reports: *Mass Spectrometry, A Review of the Recent Literature Published between July 19XX and June 19XX+2*, Vol. 1 (1968–1970), 1971; Vol. 2 (1970–1972), 1973, Williams, DH, Ed.; Vol. 3 (1972–1974), 1975; Vol. 4 (1974–1976), 1977; Vol. 5 (1976–1978), 1979; Vol. 6 (1978–1980), 1981; Vol. 7 (1980–1982), 1984, Johnstone, RAW, Ed.; Vol. 8 (1982–1984), 1985; Vol. 9 (1984–1986), 1987; Vol.10 (1986–1988), 1989, Rose, ME, Ed.; Royal Society of Chemistry (formerly Chemical Society): Cambridge, U.K.

8. Ogata, K; Hayakawa, T, Eds. *Recent Developments in Mass Spectroscopy*, Proceedings of the International Conference on Mass Spectroscopy, Kyoto, Japan, September 8–12, 1969; University of Tokyo Press: Tokyo, 1970.

9. Price, D; Williams, JE, Eds. *Time of Flight Mass Spectrometry* (1st proceeding), 1969; all subsequent proceedings entitled *Dynamic Mass Spectrometry*, Vol. 1., 1970 (2nd); Vol. 2, 1971 (3rd); Vol. 3, 1972 (4th) Price, D, Ed.; Vol. 4, 1976 (5th), Price, D; Todd, JFJ, Eds.; Vol. 5, 1978 (6th); Vol. 6, 1981 (7th); published by Heyden: London as the proceedings of the seven European Time-of-Flight Symposia.

10. Massey, HSW; Burhop, EHS; Gilbody, HB *Electronic and Ionic Impact Phenomena*, 2nd ed., Vol. I *Electron Collisions with Atoms*, 1969; Vol. II *Electron Collisions with Molecules – Photoionization*, 1969; Vol. III *Slow Collisions of Heavy Particles*, 1971; Vol. IV *Recombination and Fast Collisions of Heavy Particles*, 1974; Vol. V *Slow Positron and Muon Collisions – Notes on Recent Advances*, 1974; Oxford University Press: London.

11. Kientiz, H *Massenspektrometrie*; Verlag Chemie: Weinheim, Germany, 1968.

12. Reed, RI, Ed. *Modern Aspects of Mass Spectrometry*, Proceedings of the 2nd NATO Advanced Study Institute of Mass Spectrometry on Theory, Design, and Applications, July 1966, University of Glasgow, Glasgow, Scotland; Plenum: New York, 1968.

13. Blauth, EW *Dynamic Mass Spectrometers* (translated from German); Elsevier: Amsterdam, 1966.

14. Jayaram, R *Mass Spectrometry: Theory and Applications*; Plenum: New York, 1966.

15. Mead, WL, Ed. *Advances in Mass Spectrometry*, Vol. 3; Pergamon: New York, 1966.

16. Thomson, GP *J. J. Thomson and the Cavendish Laboratory in His Day*; Doubleday: New York, 1965.

17. Reed, RI, Ed. *Mass Spectrometry*, Proceedings of the 1st NATO Advanced Study Institute of Mass Spectrometry on Theory, Design, and Applications; Academic: London, 1965.

18. McDaniel, EW *Collision Phenomena in Ionized Gases*; Wiley: New York, 1964.

19. McDowell, CA, Ed. *Mass Spectrometry*; McGraw-Hill: New York, 1963. (reprinted by Robert E. Krieger: Huntington, NY, 1979)

20. Elliot, RM, Ed. *Advances in Mass Spectrometry*, Vol. 2; Pergamon: New York, 1963.

21. Waldron, JD, Ed. *Advances in Mass Spectrometry*, Vol. 1; Pergamon: New York, 1959.

22. Duckworth, HE *Mass Spectroscopy*; Cambridge: London, 1958.

23. Rieck, GR *Einführung in die Massenspektroskopie* (translated from Russian); VEB Deutscher Verlag der Wissebschaften: Berlin, 1956.

24. Loeb, LB *Basic Processes of Gaseous Electronics*; University of California Press: Berkeley, CA, 1955. (reprinted in 1960 with Appendix I)

25. Robertson, AJB *Mass Spectrometry: Methuen's Monographs on Chemical Subjects*; Wiley: New York, 1954.

26. Blears, J (Chairman, Mass Spectrometry Panel, Institute of Petroleum) *Applied Mass Spectrometry*, a report of a conference organized by The Mass Spectrometry Panel of The Institute of Petroleum, London, 29–31 October 1953; The Institute of Petroleum: London, 1954.

27. Barnard, GP *Modern Mass Spectrometry*; American Institute of Physics: London, 1953.

28. Hipple, JA; Aldrich, LT; Nier, AOC; Dibeler, VH; Mohler, FL; O'Dette, RE; Odishaw, H; Sommer, H (Mass Spectroscopy Committee) *Mass Spectrometry in Physics Research*; National Bureau of Standards Circular 522, United States Government Printing Office: Washington, DC, 1953.

29. Ewald, H; Hintenberger, H *Methoden und Anwendunyngen der Massenspektroskopie*; Verlag Chemie: Weinheim, Germany, 1952; English translation by USAEC, Translation Series AEC-tr-5080; Office of Technical Service: Washington, DC, 1962.

30. Massey, HSW; Burhop EHS *Electronic and Ionic Impact Phenomena*; Oxford University Press: London, 1952. (2nd printing, 1956)

31. Massey, HSW *Negative Ions*, 2nd ed.; Cambridge at the University: London, 1950. (1st ed., 1933)

32. Mott, NF; Massey, HSW *The Theory of Atomic Collisions*, 2nd ed.; Oxford University Press: London, 1949.

33. Aston, FW *Mass Spectrometry and Isotopes*, 2nd ed.; Edward Arnold: London, 1942. (1st ed., 1933)

34. Aston, FW *Isotopes*, 2nd ed.; Edward Arnold: London, 1924. (1st ed., 1922)

35. Thomson, JJ *Rays of Positive Electricity and their Application to Chemical Analysis*, 2nd ed.; Longmans Green: London, 1921. (1st ed., 1913)

Collections of Mass Spectra in Hardcopy

There have been many collections of mass spectra that have come and gone. In a 1985 monograph (A Guide To, And Commentary On, The Published Collection and Literature of Mass Spectral Data) published by VG Analytical (the United Kingdom mass spectrometry company now known as Micromass/Waters), 33 separate collections were referenced. In 1974 and 1978, the American Society for Mass Spectrometry published the 1st and 2nd editions of *A Guide to Collections of Mass Spectral Data*. These editions include 24 and 30 references, respectively. None of these collections has been lost. They have all been consolidated into either the National Institute of Standards and Technology NIST98 Mass Spectral Database or the Wiley Registry of Mass Spectral Data, or both. Some of these collections are not currently available in an electronic format, or the electronic format is only of abbreviated spectra that ranges from a minimum of 16 to a maximum of 50 mass spectral peaks (Collections 6). Hardcopy volumes are somewhat less valuable than electronic versions. The Cornu collection is even less valuable because it is a tabular listing of the 10 most intense peaks.

1. Makin, HLJ; Trafford, DJH; Nolan, J *Mass Spectra and GC Data of Steroids: Androgens and Estrogens*; Wiley: Chichester, U.K., 1999.

2. Newman, R; Gilbert, MW; Lothridge, K *GC-MS Guide to Ignitable Liquids*; CRC: Boca Raton, FL, 1998.

3. ‡Vickerman, JC; Briggs, D; Henderson, A, Eds. *The Wiley Static SIMS Library*; Wiley: New York, 1996.

4. ‡Adams, RP *Identification of Essential Oil Compounds by Gas Chromatography/Mass Spectrometry*; Allured: Carol Stream, IL, 1995. (1,211 spectra)

5. ‡Mills, T, III; Roberson, JC *Instrumental Data For Drug Analysis*, 2nd ed., Vols. 1–5, Vol. 6, Vol. 7; CRC: Boca Raton, FL, 1993. (originally published by Elsevier: New York, 1987–1992)

6. ‡Pfleger, K; Maurer, WW; Weber, A *Mass Spectral and GC Data of Drugs, Pollutants, Pesticides and Metabolites*, 2nd ed., 3-volume set; VCH: New York, 1992.

7. Hites, RA *CRC Handbook of Mass Spectra of Environmental Contaminants*, 2nd ed.; CRC: Boca Raton, FL, 1992. (533 spectra)

8. McLafferty, FW; Stauffer, DB *Important Peak Index of the Registry of Mass Spectral Data*, 3 volumes; Wiley: New York, 1991.

9. *The Eight Peak Index of Mass Spectra*, 4th ed.; Royal Society of Chemistry: Cambridge, U.K., 1991.

10. ‡McLafferty, FW; Stauffer, DB *The Wiley/NBS Registry of Mass Spectral Data*, 7 volumes; Wiley: New York, 1989. (133,000 spectra)

11. Stemmler, EA; Hites, RA *Electron Capture Negative Ion Mass Spectra of Environmental Contaminants and Related Compounds*; VCH: New York, 1988. (361 spectra)

12. †Heller, SR; Milne, GWA; Gevantman, LH *EPA/NIH Mass Spectral Data Base, Supplement 2, 1983*; National Standard Reference Data System, National Bureau of Standards, Department of Commerce, United States Government, 1983. (6,557 spectra)

13. †Sunshine et al. *CRC Handbook of Mass Spectra of Drugs*; CRC: Boca Raton, FL, 1981. (1,208 EI spectra and 628 CI spectra)

14. †Middleditch, BS; Missler, SR; Hines, HB *Mass Spectrometry of Priority Pollutants*; Plenum: New York, 1981. (114 spectra)

15. †Heller, SR; Milne, GWA *EPA/NIH Mass Spectral Data Base, Supplement 1, 1980*; National Standard Reference Data System, National Bureau of Standards, Department of Commerce, United States Government, 1980. (8,807 spectra)

16. †‡Heller, SR; Milne, GWA *EPA/NIH Mass Spectral Data Base*, 4 volumes and an index; National Standard Reference Data System, National Bureau of Standards, Department of Commerce, United States Government, 1978. (23,556 spectra)

17. †Cornu, A; Massot, R *Compilation of Mass Spectral Data*, 2nd ed., 2 volumes; Heyden: Philadelphia, PA, 1975. (10,000 spectra)

18. †Safe, S; Hutzinger, O *Mass Spectrometry of Pesticides and Pollutants*; CRC: Cleveland, OH, 1973. (275 spectra)

‡ Also available in electronic format
† Out of print

Mass Spectrometry, GC/MS, and LC/MS Journals

Articles containing information on mass spectrometry can be found in many different scientific journals as well as those listed below. This list consists of journals that are specific to mass spectrometry (Journals 1, 2, 3, 4, 5, 6, 8, 9, 15, and 16), that pertain to a specific analytical technique (Journals 10 and 11), or that pertain to general chemistry (Journals 7, 12, 13, and 14). Some of the journals have complementary subscriptions (Journals 12, 13, and 14), whereas other journals have annual subscription rates of thousands of dollars (Journals 1, 4, 6, 10, and 11). Some journals have reasonable society membership rates (Journals 2 and 7). The more expensive journals will often have reasonable individual subscription prices (Journals 1 and 4).

In addition to review and research articles, most of these journals also provide reviews of software, books, and other items of interest to the mass spectrometrist. The exceptions are the proceedings of meetings (Journals 16 and 17) and list and/or abstract sources (Journals 8 and 9).

One of the interesting features of *JMS* (Journals 1) is a section entitled "Current Literature in Mass Spectrometry" that appears at the end of every issue. This feature is a bibliography of articles published over the past six to eight weeks. It is divided into 11 major sections with the Biology/Biochemistry section subdivided into 4 additional categories. At the end of each volume, all the listings for the year are made available in a Microsoft® Access format that can be searched electronically.

1. *Journal of Mass Spectrometry*; Wiley: New York.
   (formerly *Organic Mass Spectrometry*, incorporating *Biomedical Mass Spectrometry*)

2. *Journal of the American Society for Mass Spectrometry*; Elsevier: New York.

3. *European Mass Spectrometry*; IM Publications: West Sussex, U.K.

4. *Rapid Communications in Mass Spectrometry*; Wiley: New York.

5. *Mass Spectrometry Reviews*; Wiley: New York.

6. *International Journal of Mass Spectrometry and Ion Processes*; Elsevier: New York.

7. *Analytical Chemistry*; American Chemical Society: Washington, DC.

8. *CA Selects Plus: Mass Spectrometry*; American Chemical Society: Washington, DC.

9. *Mass Spectrometry Bulletin*; Royal Society of Chemistry: Cambridge, U.K.

10. *Journal of Chromatography A*; Elsevier: New York.

11. *Journal of Chromatography B*; Elsevier: New York.

12. *American Laboratory*; ISC: Shelton, CT.

13. *LC/GC*; Advanstar: Eugene, OR.

14. *Spectroscopy*; Advanstar: Eugene, OR.

15. *Proceedings of the nth ASMS Conference on Mass Spectrometry and Allied Topics*, published annually; American Society for Mass Spectrometry: Santa Fe, NM (ASTM E14 Committee Meetings began in 1952; published annually from 1961–1969 ASTM E14 Committee Meeting Proceedings; ASMS began in 1970 with the 18th Conference Proceedings).

16. *Advances in Mass Spectrometry*, Proceedings of the Triennial International Mass Spectrometry Conference. Vols. 1, 2, and 3 (1958, 1961, and 1964) are found in the **Books of Historical Significance** section of this bibliography. Vols. 12, 13, and 14 (1991, 1994, and 1997) are found in the **Reference Books** section of this bibliography. There are Vols. 4–11 for meeting years 1967, 1970, 1973, 1976, 1979, 1982, 1985, and 1988. The first six meetings were held in the United Kingdom. Beginning with the 7th meeting (Florence, Italy), each succeeding meeting has been held in a different European country. The next meeting is scheduled for Barcelona, Spain, in August 2000.

Personal Computer MS Abstract Sources

1. *Current Contents*® *on CD-ROM, Physical, Chemical & Earth Sciences*; Institute for Scientific Information: Philadelphia, PA; FREE Demo available.

2. *Analytical Abstracts on CD-ROM*; Royal Society of Chemistry: Cambridge, U.K.; or SilverPlatter Information: Norwood, MA; FREE 30-day Trial Subscription.

3. *CASurveyor: Mass Spectrometry and Applications*; Chemical Abstracts Service (American Chemical Society): Washington, DC; FREE Demo available.

4. *LC/MS Update* (1991–current); *GC/MS Update Part A: Environmental* (1991–1996); and *GC/MS Update Part B: Biomedical, Clinical, Drugs* includes forensics (1991–current); HD Science: Newport, Wilmington, DE; or HD Science Limited: Nottingham, U.K.

5. *The PC Version of the Mass Spectrometry Bulletin*; Royal Society of Chemistry: Cambridge, U.K.

6. Annual Collections of the "Current Literature in Mass Spectrometry" in the *Journal of Mass Spectrometry* (1995–1998); Wiley: Chichester, U.K.

Integrated Spectral Interpretation Books

General integrated spectral interpretation books include information on the interpretation of proton NMR, IR, and mass spectra as well as how to use ultraviolet data in conjunction with these three spectral techniques. Some books also include a section on ^{13}C NMR. These books are good for an overview of the subject, but do not provide the in-depth mass spectrometry interpretational information.

In addition to the books listed below, the Mathieson book (Interpretation 22) is also an integrated book in that it includes separate sections on NMR and IR as well as the one on mass spectrometry.

1. Crews, P; Jaspars, M; Rodriquez, J *Organic Structure Analysis*, 1st ed.; Oxford University Press: Oxford, U.K., 1998.

2. Lambert, JB; Shurvell, HF; Lightner, DA; Cooks, RG *Organic Structural Spectroscopy*; Prentice Hall: Upper Saddle River, NJ, 1998.

3. Silverstein, RM; Wobster, FX *Spectrometric Identification of Organic Compounds*, 6th ed.; Wiley: New York, 1998. (1st ed., 1963; 2nd ed., 1967, Silverstein, RM; Bassler, GC; 3rd ed., 1974; 4th ed., 1981; 5th ed., 1991, Silverstein, RM; Bassler, GC; Morrill, TC)

4. Harwood, LM; Claridge, TDW *Introduction to Organic Spectroscopy*, 1st ed.; Oxford University Press: New York, 1997.

5. Hesse, M; Meier, H; Zeeh, B (Linden, A; Murray, M, Translators) *Spectroscopic Methods in Organic Chemistry*, 1st ed.; Thieme: New York, 1997.

6. Pavia, DL; Lampman, GM; Kriz, GS *Introduction to Spectroscopy: A Guide for Students of Organic Chemistry*, 2nd ed.; Saunders College: Orlando, FL, 1996. (1st ed., 1979)

7. Field, LD; Sternhell, S; Kalman, JR *Organic Structures from Spectra*, 2nd ed.; Wiley: Chichester, U.K., 1995. (1st ed., 1986)

8. Feinstein, K *Guide to Spectroscopic Identification of Organic Compounds*; CRC: Boca Raton, FL, 1995.

9. Williams, DH; Fleming, I *Spectroscopic Methods in Organic Chemistry*, 5th ed.; McGraw-Hill: London, 1995. (1st ed., 1966)

10. Jones, C; Mulloy, B; Thomas, AH, Eds. *Spectroscopic Methods and Analyses (NMR, Mass Spectrometry, and Metalloprotein Techniques)*, Vol. 17 *Methods in Molecular Biology*; Humana: Totowa, NJ, 1993.

11. Kemp, W *Organic Spectroscopy*, 3rd ed.; W. H. Freeman: New York, 1991.

12. Fresenius, W; Huber, JFK; Pungor, E; Rechnitz, GA; Simon, W; West, TS, Eds. *Tables of Spectral Data for Structure Determination of Organic Compounds*, 2nd English ed., translated from the German edition by K. Biemann; Springer-Verlag: Berlin, Germany, 1989.

13. Sorrell, TN *Interpreting Spectra of Organic Molecules*; University Science: Mill Valley, CA, 1988.

14. Scheinmann, F, Ed. *An Introduction to Spectroscopic Methods for the Identification of Organic Compounds*, Vol. 2 *Mass Spectrometry, Ultraviolet Spectroscopy, Electron Spin Resonance Spectroscopy, NMR (Recent Developments), Use of Various Spectral Methods Together, and Documentation of Molecular Spectra*; Pergamon: Oxford, U.K., 1974.

15. Mathieson, DW, Ed. *Interpretation of Organic Spectra*; Academic: New York, 1965.

Monographs

All the citations in this segment are from VG Instruments/Micromass. All instrument manufacturers publish application notes; however, Micromass (and its preceding companies) is the only manufacturer that has published this type of general-topic monograph. These monographs are like review articles found in *Mass Spectrometry Reviews* or the Special Features section of the *Journal of Mass Spectrometry*. These monographs are not as well referenced as the articles in these two journals, but do provide a good overview of the subject. Such promotional material is of benefit to those wanting to get a quick understanding of a topic, and it is hoped that more of this type of material will be forthcoming.

Another example of the ready-reference-material approach to information dissemination is found in the Siuzdak book (Introductory 6). This book was written to provide a quick understanding of mass spectrometry to biotechnology executives who have to make financial decisions about mass spectrometry instrumentation and facilities.

1. Rose, ME *Modern Practice of Gas Chromatography/Mass Spectrometry*, VG Monographs in Mass Spectrometry, No. 1; VG Instruments: Altrincham, U.K.

2. Mellon, FA *Liquid Chromatography/Mass Spectrometry*, VG Monographs in Mass Spectrometry, No. 2; VG Instruments: Altrincham, U.K.

3. Clench, MR *A Comparison of Thermospray, Plasmaspray, Electrospray and Dynamic FAB*, VG Monographs in Mass Spectrometry, No. 3; VG Instruments: Altrincham, U.K.

4. Scrivens, JH; Rollins, K *Tandem Mass Spectrometry*, VG Monographs in Mass Spectrometry, No. 4; VG Analytical, Fisons Instruments: Altrincham, U.K.

5. Ardrey, B *Mass Spectrometry in the Forensic Sciences*, VG Monographs in Mass Spectrometry, No. 5; VG Analytical, Fisons Instruments: Altrincham, U.K.

6. Hsu, J-P *High and Low Resolution GC/MS in Environmental Sciences*, VG Monographs in Mass Spectrometry, No. 6; VG Analytical, Fisons Instruments: Altrincham, U.K.

Software

There are several programs that are available as self-training. Those programs developed by the U.K. company, Cognitive Solutions (Software 10–13), are somewhat like English roast beef. They are intellectually nutritious; however, they fail to excite the experiential palate. Equivalent titles available from SAVANT (Software 14–22) will hold the user's interest in a much more conductive manner for learning. Just as the *MS Fundamentals* program is a seminal tool in the development of the understanding of the transmission-quadrupole mass spectrometer, *Fundamentals of GC/MS* is one of the better instrument-user software packages developed. This program was initially developed and sent out several times for review to a number of people who are involved in training on various aspects of GC/MS and in the development of training programs. The result is what is assured to become an award-winning effort.

All of the training programs from SAVANT and Cognitive Solutions were developed in Interactive ToolBook, a powerful tool for the development of training programs. All have tests built into the programs to allow the user to evaluate the results of the training.

The two volumes of *SpectraBook* are also based on the Interactive ToolBook platform. These two programs each contain data on 50 separate compounds: mass, proton NMR, ^{13}C NMR and infrared spectra as well as the structure, molecular weight based on the atomic weights of each compound's elements, physical properties, and several synonyms. Help files are provided to assist the user in developing desired interpretational skills. Another nice feature is the ability to display which properties result in specific spectral peaks. Placing the Mouse pointed on a labeled mass spectral peak and holding down the left Mouse button will result in a display of the mechanism(s) that produced the ion represented by that peak. Similar displays are provided for the other types of spectra.

It is unfortunate that the author of *SpectraBook* (programs copyrighted in 1990 and 1992) did not take more care to be correct in some of the presentations such as the use of *m/z* as the symbol for mass-to-charge (inappropriately written Mass to Charge on the abscissa of mass spectra) ratio instead of the m/e symbol, which was replaced in the 1970s. The indicated shift of pairs of electrons in the displayed mechanism for beta cleavage resulting from a gamma-hydrogen shift does not instill confidence in the accuracy of instruction.

The programs published by ChemSW (Software 1–9) are very well thought out and provide utilities that are not found in the data-system software for most, if not all, commercially available instruments. The titles of these programs are self-explanatory.

The two programs associated with the Wiley and NIST Mass Spectral Databases (Software 31 and 33) are widely available from a number of different sources. A number of GC/MS and LC/MS programs now provide the NIST Mass Spectral Search Program as the search routine used with their proprietary instrument software. Both of the programs are capable of reading most, if not all, commercially available instrument data formats.

1. *CESAR*™ Capillary Electrophoresis Simulation for Application Research; ChemSW: Fairfield, CA.

2. *GC-SOS*™ Gas Chromatography Simulation and Operation Software, Ver. 5; ChemSW: Fairfield, CA.

3. Bernert, JT, Jr. (Quadtech Associates) *Mass Spec Calculator*™ Ver. 3, and *Mass Spec Calculator*™ *Pro*; ChemSW: Fairfield, CA.

4. Junk, T *GC and GC/MS File Translator*™ *Professional*; ChemSW: Fairfield, CA.

5. Junk, T *GC and GC/MS File Manager*™; ChemSW: Fairfield, CA.

6. Bernert, JT, Jr. (Quadtech Associates) *Mass Differential Analysis Tools*; ChemSW: Fairfield, CA.

7. *Protein Tools*™; ChemSW: Fairfield, CA.

8. *HPLC Optimization*™; ChemSW: Fairfield, CA.

9. *GPMAW*™ General Protein Mass Analysis for Windows; ChemSW: Fairfield, CA.

10. *Interactive Training Program™ *Gas Chromatography*; published by Cognitive Solutions: Glasgow, U.K., a.k.a. Softbooks out of the United States; distributed by ChemSW: Fairfield, CA.

11. ‡Interactive Training Program™ *High Performance Liquid Chromatography*; published by Cognitive Solutions: Glasgow, U.K., a.k.a. Softbooks out of the United States; distributed by ChemSW: Fairfield, CA.

12. ‡Interactive Training Program™ *Advanced Gas Chromatography*; published by Cognitive Solutions: Glasgow, U.K., a.k.a. Softbooks out of the United States; distributed by ChemSW: Fairfield, CA.

13. ‡Davis, S (HD Technologies) Interactive Training Program™ *Mass Spectrometry*; published by Cognitive Solutions: Glasgow, U.K., a.k.a. Softbooks out of the United States; distributed by ChemSW: Fairfield, CA.

14. ‡Saunders, D *Introduction to Gas Chromatography*; SAVANT:† Fullerton, CA.

15. ‡Saunders, D *Fundamentals of Gas Chromatography/Mass Spectrometry*; SAVANT: Fullerton, CA.

16. ‡*Introduction to High Performance Liquid Chromatography*; SAVANT: Fullerton, CA.

17. ‡*Method Development in High Performance Liquid Chromatography*; SAVANT: Fullerton, CA.

18. ‡*High Performance Liquid Chromatography Equipment*; SAVANT: Fullerton, CA.

19. ‡*Troubleshooting High Performance Liquid Chromatography*; SAVANT: Fullerton, CA.

20. ‡*Separation Modes of High Performance Liquid Chromatography*; SAVANT: Fullerton, CA.

21. ‡*HPLC Calculation Assistant & Reference Tables*; SAVANT: Fullerton, CA.

22. ‡*Identification & Quantification for HPLC*; SAVANT: Fullerton, CA.

23. ‡Hart, M *MS Fundamentals*; published by Hewlett-Packard: Palo Alto, CA; distributed by SAVANT: Fullerton, CA.

24. ‡Schatz, PF *SpectraBook*, Vol. 1, 1990 and Vol. 2, 1992; published by Falcon Software; distributed by SAVANT: Fullerton, CA.

25. *DryLab*; LC Resources: Lafayette, CA.

26. *Introduction to CE*; LC Resources: Lafayette, CA.

27. Figueras, J *Mass Spec*, Ver. 3.0; Trinity Software: Plymouth, NH.

28. *ACD/SpecManager: MS Module*; Advanced Chemistry Development: Toronto, Canada.

29. *Mass Frontier*, ThemoQuest/HighChem: San Jose, CA.

30. *MASSTransit*; Palisade: Newfield, NY.

31. *Benchtop PBM* with *Wiley Registry of Mass Spectral Data*, Ver. 6 or 6N, or Select; Palisade: Newfield, NY.

32. Dahl, D *SIMION 3D*, Ver. 6.0, Ion and Electron Optics Program, Scientific Instrument Services: Ringoes, NJ.

33. *NIST Mass Spectral Search Program for Windows*, Ver. 1.7 and *NIST/EPA/NIH Mass Spectral Database* with *AMDIS*, Ver. 2.0, Automated Mass Spectral Deconvolution and Identification System, National Standard Reference Data System, National Institute of Standards and Technology: Gaithersburg, MD.

‡ Uses Asymetrix ToolBook Runtime
† Sloane Audio Visuals for Analysis and Training

WEB SITES FOR SOFTWARE COMPANIES:

ChemSW	http://www.chemsw.com
LC Resources	http://www.lcresources.com
NIST	http://www.nist.gov/srd/analy.htm
Palisade	http://www.palisade.com
Scientific Instrument Services	http://www.sisweb.com
Trinity Software	http://www.trinitysoftware.com
Advanced Chemistry Development	http://www.acdlabs.com
HighChem	http://www.highchem.com

INDEX

A

A.M.U., a.m.u., 34, 35
Abbreviations Usage, 53
ablation, 52
Abstract Sources (PC MS), 94
accelerating voltage alternation (AVA), 19
accuracy, 41
accurate mass, 36
acylium ion, 24
alkyl ion, 24
allyl ion, 24
alpha cleavage, 20, 54
alpha-, or α-, etc. usuage, 54
American Society for Mass Spectrometry (ASMS), 1
amu, 34, 35
analyte, 44
APCI, *see* atmospheric pressure chemical ionization
API, *see* atmospheric pressure ionization
appearance energy, 24
appearance potential, 24
ASMS Guidelines, 1, 2
atmospheric pressure chemical ionization (APCI), 7
atmospheric pressure ionization (API), 7
atom, 34
atomic number, 35
atomic weight, 34, 35
AVA, *see* accelerating voltage alternation
average atomic weight, 38
average mass, 35

B

background, 42, 56
background-subtracted spectrum, 42
background subtraction, 42
base peak, 3
baseline offset, 56
benzyl ion, 24
benzylic cleavage, 20
Bibliography, 71
Biemann search, 62
Brubaker prefilter, 6

C

CAD, *see* collisionally activated dissociation
calculated exact mass, 35, 36
calibration, 39
carbanion, 25
carbenium ion, 24
carbocation, 25
carbonium ion, 24, 25
cationization, 30
centimeters-of-mercury pressure, 43
charge exchange, 18
charge inversion, 18
charge permutations, 18
charge-reduction electrospray mass spectrometry (CREMS), 9
charge-site-driven cleavage, 21
charge stripping, 18
charge transfer, 10
chemical ionization (CI), 9, 10
chemical reaction interface mass spectrometry (CRIMS), 11
chromatogram, 3
CI, *see* chemical ionization
CID, *see* collision-induced dissociation
cluster ion, 25
coaxial reflectron, 51
Collections of Mass Spectra in Hardcopy, 92
collisionally activated dissociation (CAD), 13
collision-induced dissociation (CID), 13
collision-stabilized complexes, 10
common-neutral-loss analysis, 16
Components of a Measurement, 56
continuous-dynode EM, 63, 64
continuous-flow FAB (CF-FAB), 13
conversion dynode, 63
CREMS, *see* charge-reduction electrospray mass spectrometry
CRIMS, *see* chemical reaction interface mass spectrometry
Current IUPAC Recommendations, 1, 2
curved-field reflectron, 51

D

dalton (Da), 34
Daly detector, 64
Data, 3
Data Acquisition Techniques and
    Ionization, 7
daughter ions, 28, 29
DCI, *see* desorption chemical
    ionization
decade, 49
deconvolution, 39
degrees of unsaturation, 31
delayed extraction, 51
desorption chemical ionization (DCI), 11
desorption/ionization (DI), 11
detection limit, 40, 56
DI, *see* desorption/ionization
direct-exposure probe, 11
direct infusion, 11
direct-insertion probe, 11
direct-liquid-inlet (DLI) probe, 9
discrete-dynode EM, 63
dissociation with rearrangement, 23
dissociative electron capture, 10
distonic ion, 25
distonic radical ion, 25
DLI, *see* direct-liquid-inlet probe
double-focusing mass spectrometer, 46
draw-out pulse, 51
dynode, 63

E

EC/NI, *see* electron capture/
    negative-ion detection
EI, *see* electron ionization
EICC, *see* extracted-ion-current
    chromatogram
EIEIO, *see* electron-induced excitation
    in organics or electron-impact
    excitation of ions from organics
electron capture/negative-ion
    detection (EC/NI), 10
electron energy, 25
electron impact, 12
electron-impact excitation of ions
    from organics (EIEIO), 22
electron-induced dissociation, 22
electron-induced excitation in
    organics (EIEIO), 22

electron ionization (EI), 12
electron multiplier, 63
electron volt (eV), 25, 43
electrospray (ES), 7
electrospray ionization (ESI), 7
eluant, 44
eluate, 44
eluent, 44
ES, *see* electrospray
ESI, *see* electrospray ionization
eV, *see* electron volt
even-electron ion (EE^+ or EE^-), 25
exact mass, 36
extracted-ion-current chromatogram
    (EICC), 3

F

FAB, *see* fast atom bombardment
FAB mass spectrometry, 12
Faraday cup, 64
fast atom bombardment (FAB), 12
FD, *see* field desorption
FI, *see* field ionization
field desorption (FD), 12
field ionization (FI), 12
Field's rule, 22
field-free region, 50
fit, 59
flow injection, 12
fluence, 52
formula weight, 45
Formulas and Equations, 57
forward geometry, 49
Fourier transform ion-cyclotron
    resonance (FTICR), 46
fragment ion (X^+, X^-, $X^{+\bullet}$, or $X^{-\bullet}$), 26
fragmentation pattern, 3
FTICR, *see* Fourier transform
    ion-cyclotron resonance

G

gamma-hydrogen shift-induced
    beta cleavage, 21
gridless reflectron, 51

H

heterolytic cleavage, 21
high-energy dynode, 63
Historical Books, 89
homolytic cleavage, 20
hydride abstraction, 10
hydrogen-deficiency equivalents, 31

I

IKES, *see* ion-kinetic-energy
    spectrometry
iminium ion, 26
immonium ion, 26
inches-of-mercury pressure, 43
INCOS, 61
inductive (i) cleavage, 21
in-source CAD, 16
Instruments, 46
Integrated Spectral Interpretation
    Books, 95
intermolecular, 44
International Union of Pure and Applied
    Chemistry (IUPAC), 1
Interpretation Books, 87
intramolecular, 44
Introduction, 1
Introductory Books, 73
ion, 26
Ion Detection, 63
ion-gate pulse, 51
ionization energy, 25
ion-kinetic-energy spectrometry
    (IKES), 50
ion-molecule reaction, 14
ion/molecule reaction, 14, 42
ion-pair formation that results from
    electron capture, 10
ions, 3
Ions, 24
IonSpray (ISP), 8
ion-trap mass spectrometer, 47, 48
isobaric, 26
isobaric ions, 26
isobaric peaks, 26
isotope cluster, 26
isotope number, 36
isotopes, 36
isotopic ion, 26
isotopic mass, 36

ISP, *see* IonSpray
IUPAC, *see* International Union of
    Pure and Applied Chemistry

J

Journals (MS, GC/MS, LC/MS), 93, 94

L

library, 39
Library Search, 59
Library Search Algorithms, 61
limit of quantitation, 56
line, 3
liquid secondary-ion mass spectrometry
    (LSIMS), 13
loss of the largest alkyl, 23
LSIMS, *see* liquid secondary-ion mass
    spectrometry

M

m/e, 28
m/z, 27
m/z analysis, 15
m/z analyzer, 15
magnetic-sector mass spectrometer,
    46, 47
MALDI, *see* matrix-assisted laser
    desorption/ionization
Mass, 34
mass analysis, 14
mass-analyzed ion-kinetic-energy
    spectrometry (MIKES), 50
mass chromatogram, 3
mass defect, 36, 38
mass discrimination, 6
mass filter, 47
mass fragmentography, 19
mass number, 36, 37
mass-selective detector (MSD), 19
mass spectra, 3
mass spectrograph, 14
mass spectrometer, 14
mass spectrometrist, 15
mass spectrometry, 14
mass spectrometry/mass spectrometry
    (MS/MS), 15

mass spectrophotometer, 15
mass spectroscopy, 14
mass spectrum, 3
MassTransit, 62
match factor, 59
matrix, 44
matrix-assisted laser
    desorption/ionization (MALDI), 13
McLafferty rearrangement, 21
mean-free path, 44
measured accurate mass, 36
metastable ion, 26
metastable ions, 50
microchannel plate, 64
microelectrospray, 8
microES, 8
MicroIonSpray, 8
MID, *see* multiple-ion detection
MIKES, *see* mass-analyzed
    ion-kinetic-energy spectrometry
moiety, 45
molecular ion ($M^{+\bullet}$ or $M^{-\bullet}$), 28
molecular mass, 35, 38
molecular weight, 35
molecules, 37
Monographs, 96
monoisotopic mass, 37
monoisotopic mass spectrum, 4
most abundant mass, 37
MRM, *see* multiple reaction monitoring
MSD, *see* mass-selective detector
MS/MS, *see* mass spectrometry/mass
    spectrometry
MS/MS spectrum, 4
MS^n spectrum, 4
multiple-charge ion, 31
multiple-charged ion, 31
multiple-ion detection (MID), 19
multiple reaction monitoring (MRM), 17
multiply charged ion, 31
multiply-charged ion, 31

N

N most int. Peaks, 62
N most sig. Peaks, 62
nanoelectrospray, 8
nanoES, 8
NanoFlow ES, 8
NanoSpray, 8

National Institute of Standards and
    Technology (NIST), 62
NCI, *see* negative-ion CI
negative ion/molecule reaction, 10
negative-charge ions, 30
negative-ion CI (NCI), 10
negative ions, 30
neutral loss, 37, 62
NIST, *see* National Institute of
    Standards and Technology
nitrogen rule, 37
noise, 56
nominal mass, 38
nonbonding electrons, 58
nucleon number, 36
nuclide, 38

O

octet rule, 19
odd-electron ion ($OE^{+\bullet}$ or $OE^{-\bullet}$), 25, 31
ortho effect, 22
orthogonal extraction, 51
oxonium ion, 28

P

PAD, *see* post-acceleration detector
parent ion ($P^{+\bullet}$), 28, 29
partial charge transfer, 18
particle beam (PB) interface, 9
pascal (PA), 43
Paul ion trap, 47
PBM, *see* Probability Based Matching
PCI, *see* positive-ion CI
PD, *see* plasma desorption
peak, 39
peak matching, 18
pherogram, 3
pi bond, 22
pi-bond electrons, 58
plasma desorption (PD), 13
positive-charge ions, 30
positive-ion CI (PCI), 9
positive ions, 30
post-acceleration detector (PAD), 63
post-source decay (PSD), 51
precision, 41
precursor ion, 29
precursor-ion analysis, 16

principal ion, 30, 31
probability, 59, 60
Probability Based Matching (PBM), 61
product ion, 29
product-ion analysis, 15, 16
profile, 3
profile data, 3
proton transfer, 10
protonated molecular ion, 30
protonated molecule, 30
PSD, *see* post-source decay
pseudo-molecular ion, 32
purity, 59

Q

quadrupole ion-trap mass
    spectrometer, 47
quadrupole mass spectrometer, 47
quality, 60
quasi-molecular ion, 32
Quistor, 47

R

radical (X·), 38
radical anion, 31
radical cation, 31
radical ion, 31
radical-site-driven cleavage, 20
RECI, *see* resonance electron capture
    ionization
reconstructed extracted-ion-current
    chromatogram (REICC), 3
reconstructed ion chromatogram
    (RIC), 3
reconstructed total-ion-current
    chromatogram (RTICC), 5
Reference Books, 75
Referenced Citations in Documents, 65
References, 66
reflectron, 52
REICC, *see* reconstructed
    extracted-ion-current chromatogram
relative atomic weight, 38
resolution, 32
resolving power, 32
resonance electron capture, 10
resonance electron capture
    ionization (RECI), 10

retro-Diels–Alder reaction, 22
reverse geometry, 49
reverse match factor, 60
reverse search, 60
Rfit (reverse fit), 60
rings plus double bonds, 31
RTI & spectrum, 62
RIC, *see* reconstructed ion
    chromatogram
RTICC, *see* reconstructed
    total-ion-current chromatogram

S

S/B, *see* signal-to-background ratio
S/N, *see* signal-to-noise ratio
sample, 44
satellite ions, 33
satellite peaks, 33
scan, 5
Search Algorithms, 61
secondary-ion mass spectrometry
    (SIMS), 19
selected ion monitoring (SIM), 19
selected reaction monitoring
    (SRM), 17
selective ion monitoring, 19
self-CI, 42, 43
sensitivity, 39, 40, 41
SID, *see* surface-induced dissociation
SI units, 43
sigma bond, 23
sigma-bond cleavage, 23
sigma-bond electrons, 58
signal strength, 56
signal-to-background ratio (S/B), 41, 56
signal-to-noise ratio (S/N), 41, 56
SIM, *see* selected ion monitoring
SIM mass chromatogram, 5
SIM plot, 5
similarity index, 60
simple cleavage, 23
SIMS, *see* secondary-ion mass
    spectrometry
skeletal rearrangements, 23
Software, 97
solids probe, 11
Some Other Important Definitions, 39
Some Other Terminology to Avoid, 45
space charge, 42
spectral skewing, 6

SRM, *see* selected reaction monitoring
Stevenson's rule, 23
surface-induced dissociation (SID), 17
Système International d'Unités, 43

T

Technique-Oriented Books, 80
Terms Associated with Double-Focusing
    Mass Spectrometers, 49
Terms Associated with Time-of-Flight
    Mass Spectrometers, 51
Terms Associated with Computerized
    Spectral Matching, 59
ThermaBeam, 9
thermospray (TSP), 9
thomson, 27, 28
TIC, *see* total-ion chromatogram
TICC, *see* total-ion-current
    chromatogram
time-of-flight mass spectrometer, 48
tolyl ion, 33
Torr, torr, 43
total-ion chromatogram (TIC), 5
total ion current, 5
total-ion-current chromatogram
    (TICC), 5

transmission-quadrupole mass
    spectrometer, 47
tropylium ion, 33
TSP, *see* thermospray
tuning, 39
Turner–Kruger ion-optics lens, 6
Types of Elements and Electrons, 58
Types of *m/z* Analyzers, 46

U

u, 34, 35
Use of Abbreviations, 53

W

whole number rule, 38

X

X elements, 58
X+1 elements, 58
X+2 elements, 58

PROOF OF A MOLECULAR ION PEAK – M$^{+\bullet}$

1. If a compound is known, the molecular ion has a mass-to-charge ratio (m/z) value equal to the sum of the atomic masses of the most abundant isotope of each element that comprises the molecule (assuming the ion is a single-charge ion).

2. The nominal molecular weight of a compound, or the m/z value for the molecular ion, is an even number for any compound containing only C, H, O, S, Si, P, and the halogens.

    Fragment ions derived via homolytic, heterolytic, or sigma-bond cleavage from these molecular ions (even m/z) have an odd m/z value and an even number of electrons.

    Fragment ions derived from these molecular ions (even m/z) via expulsion of neutral components (e.g., H_2O, CO, ethylene, etc.) have an even m/z value and an odd number of electrons.

3. **Nitrogen rule**: A compound containing an odd number of nitrogen atoms—in addition to C, H, O, S, Si, P, and the halogens—has an odd molecular weight.

    Molecular ions of these compounds fragment via homolytic, heterolytic, or sigma-bond cleavage to produce ions of an even m/z value unless the nitrogen atom is lost with the neutral radical.

    An even number of nitrogen atoms in a compound results in an even nominal molecular weight.

4. The molecular ion peak must be the highest m/z value of any significant (nonisotope or nonbackground) peak in the spectrum. Corollary: The highest m/z value peak observed in the mass spectrum need not represent a molecular ion.

5. The peak at the next lowest m/z value in the mass spectrum must not correspond to the loss of an impossible or improbable combination of atoms.

6. No fragment ion may contain a larger number of atoms of any particular element than the molecular ion.

Courses on the interpretation of mass spectra and techniques of mass spectrometry are offered by the Continuing Education Department of the American Chemical Society (http://www.acs.org/education/profdevl/short.html).

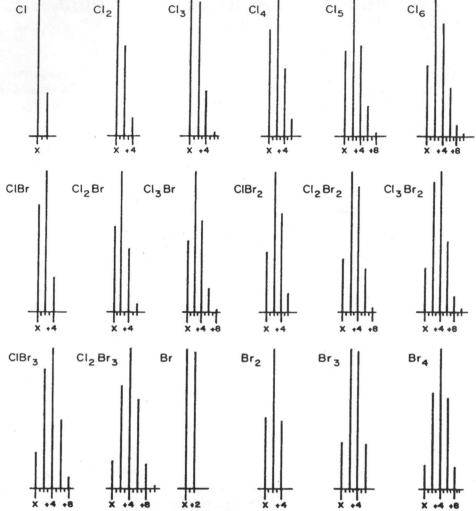

Graphical representation of relative isotope peak intensities for any given ion containing the indicated number of chlorine and/or bromine atoms. Numeric values are on back cover.

Steps to Determine Elemental Composition Based on Isotope Peak Ratios

1. Determine the nominal m/z value peak (peak at lowest m/z value, above which other peaks can be attributed to isotopic multiplicity or background).
2. Assign the X+2 elements, except oxygen.
3. Assign the X+1 elements. (Remember to normalize X+1 to X, if necessary.)
4. Balance the mass.
5. Assign the atoms of oxygen.
6. Balance the mass.
7. Assign the X elements.
8. From the empirical formula, determine the number of rings plus double bonds.
9. Propose a possible structure.
10. Does it make sense?

Atoms of ClBr	X	X+2	X+4	X+6	X+8	X+10
Cl	100	32.5				
Cl_2	100	65.0	10.6			
Cl_3	100	97.5	31.7	3.4		
Cl_4	76.9	100	48.7	0.5	0.9	
Cl_5	61.5	100	65.0	21.1	3.4	0.2
Cl_6	51.2	100	81.2	35.2	8.5	1.1
ClBr	76.6	100	24.4			
Cl_2Br	61.4	100	45.6	6.6		
Cl_3Br	51.2	100	65.0	17.6	1.7	
$ClBr_2$	43.8	100	69.9	13.7		
Cl_2Br_2	38.3	100	89.7	31.9	3.9	
Cl_3Br_2	31.3	92.0	100	49.9	11.6	1.0
$ClBr_3$	26.1	85.1	100	48.9	8.0	
Cl_2Br_3	20.4	73.3	100	63.8	18.7	2.0
Br	100	98.0				
Br_2	51.0	100	49.0			
Br_3	34.0	100	98.0	32.0		
Br_4	17.4	68.0	100	65.3	16.0	

Chlorine and Bromine Isotopic Abundance Ratios

M - 1	loss of hydrogen radical	M - $^{\bullet}$H
M - 15	loss of methyl radical	M - $^{\bullet}CH_3$
M - 29	loss of ethyl radical	M - $^{\bullet}CH_2CH_3$
M - 31	loss of methoxyl radical	M - $^{\bullet}OCH_3$
M - 43	loss of propyl	M - $^{\bullet}CH_2CH_2CH_3$
M - 45	loss of ethoxyl	M - $^{\bullet}OCH_2CH_3$
M - 57	loss of butyl radical	M - $^{\bullet}CH_2CH_2CH_2CH_3$
M - 2	loss of hydrogen	M - H_2
M - 18	loss of water	M - H_2O
M - 28	loss of CO or ethylene	M - CO or M - C_2H_4
M - 32	loss of methanol	M - CH_3OH
M - 44	loss of CO_2	M - CO_2
M - 60	loss of acetic acid	M - CH_3CO_2H
M - 90	loss of silanol: $HO-Si(CH_3)_3$	M - $HO-SI-(CH_3)_3$

Common Neutral Losses

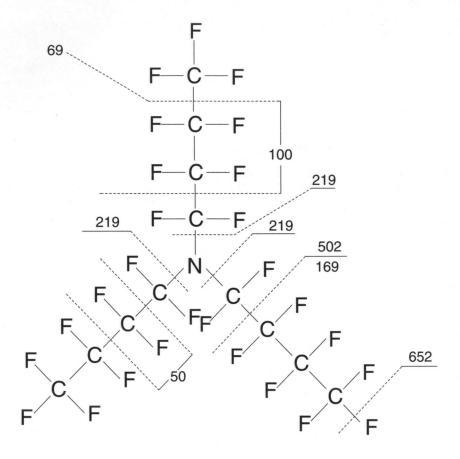

The molecular weight of perfluorotributylamine (PFTBA, a.k.a. FC-43), used to calibrate the *m/z* scale of transmission-quadrupole and quadrupole ion-trap mass spectrometers operated in the electron ionization mode, is 671. The following is an explanation of the origin of some of the peaks observed in its EI mass spectrum:

671			671			671	
$-\underline{207}$	(3×69)		$-\underline{507}$	(3×169)		$-\underline{57}$	(3×19)
464			164			614	
$\underline{-50}$			$+\underline{100}$				
414			264			169	
						$-\underline{38}$	(2×19)
						131	

Other Books From Global View Publishing

A Global View of LC/MS: How to solve your most challenging analytical problems.

Ross Willoughby, Ed Sheehan, and Sam Mitrovich

This text is a definitive problem-solving guide to the rapidly expanding field of liquid chromatography/mass spectrometry. A structured approach is utilized throughout the text to evaluate and solve many of the problems faced in the analysis of complex and labile samples. The reader is presented with techniques to evaluate and select the many alternative technologies that are available today with LC/MS. You will learn to:

Evaluate the various LC/MS technologies	Maintain an effective LC/MS laboratory
Evaluate your needs for LC/MS	Solve real analytical problems
Acquire LC/MS	Develop an individualized plan
Set up an effective LC/MS laboratory	Develop and validate LC/MS methods

Published 1998 – First edition (Second edition available in 2001)   ISBN 0-9660813-0-7

A Global View of MS/MS: How to solve problems with the ultimate analytical tool.

Ross Willoughby and Ed Sheehan

This text is a comprehensive guide to the increasingly applicable technique of mass spectrometry/mass spectrometry as a tool for acquiring the four S's; namely, sensitivity, selectivity, specificity, and speed. The availability of reliable and affordable MS/MS instrumentation provides the analyst with unprecedented capabilities to solve complex analytical problems. You will learn to:

Define your specific needs for MS/MS	Interpret MS/MS spectra (CID)
Evaluate the various MS/MS technologies	Select the appropriate standardization
Acquire MS/MS capabilities	Develop and validate MS/MS methods
Select the appropriate MS/MS experiment	Acquire high throughput and productivity

Published 2000 – First edition   ISBN 0-9660813-1-5

A Global View of GC/MS: How to solve complex analytical problems in the gas phase.

O. David Sparkman, Ross Willoughby, and Ed Sheehan

This guide addresses the state-of-the-art utility of gas chromatography/mass spectrometry to solve problems with separation and detection of complex chemical species. This guidebook presents the field of GC/MS in a practical and structured problem-solving format. You will learn to:

Evaluate appropriate GC/MS technologies	Apply the latest innovations
Maximize your GC/MS performance	Minimize downtime and rework
Increase throughput	Interpret spectral/chromatographic data
Expand your analytical range	Effectively utilize database searching

Available 2001 – First edition   ISBN 0-9660813-5-8

Order information is provided on the next page.

Order Form

🖷 **FAX Orders:**   1-412-963-6882

☎ **Telephone Order:**   1-412-963-6881
                             1-888-LCMS.COM
(Have your AMEX, Discover, VISA, or MasterCard ready)

🖳 **On-line Orders:**   E-mail: success@LCMS.com
                             Internet: www.LCMS.com

🖅 **Postal Orders:**   Global View Publishing
                             P. O. Box 111384
                             Pittsburgh, PA 15238

Please Send:
    __ Copies of **Mass Spec Desk Reference** at $29.95 each
    __ Copies of **A Global View of LC/MS** at $49.95 each
    __ Copies of **A Global View of MS/MS** at $49.95 each

I understand that I may return any books for a full refund for any reason, no questions asked.

Name: _____

Company Name: _____

Address: _____

City: _____ State: _____ Zip: _____ - _____

Telephone, U.S./Canada: 1- (____)_____

Tele., Foreign: Country Code: ____ No. _____

Shipping & Handling:
USA, Ground   (3–10 days) $ 6.00
USA, Priority   (2–5 days) $12.00
International   (3–10 days) $12.00 (Global Priority or Air)

Payment:
❐ Check (Payable to Global View Publishing)
❐ Credit Card: ❐ VISA ❐ MasterCard ❐ AMEX ❐ Discover

Card Number: _____

Name on Card: _____ Exp. Date: _____/_____